CAMBRIDGE LIBRARY COLLECTION

Books of enduring scholarly value

Mathematical Sciences

From its pre-historic roots in simple counting to the algorithms powering modern desktop computers, from the genius of Archimedes to the genius of Einstein, advances in mathematical understanding and numerical techniques have been directly responsible for creating the modern world as we know it. This series will provide a library of the most influential publications and writers on mathematics in its broadest sense. As such, it will show not only the deep roots from which modern science and technology have grown, but also the astonishing breadth of application of mathematical techniques in the humanities and social sciences, and in everyday life.

Euclid and His Modern Rivals

Euclid and His Modern Rivals is a deeply convincing testament to the Greek mathematician's teachings of elementary geometry. Published in 1879, it is humorously constructed and written by Charles Dodgson (better known outside the mathematical world as Lewis Carroll, the author of Alice in Wonderland) in the form of an intentionally unscientific dramatic comedy. Dodgson, mathematical lecturer at Christ Church, Oxford, sets out to provide evidentiary support for the claim that The Manual of Euclid is essentially the defining and exclusive textbook to be used for teaching elementary geometry. Euclid's sequence and numbering of propositions and his treatment of parallels, states Dodgson, make convincing arguments that the Greek scholar's text stands alone in the field of mathematics. The author pointedly recognises the abundance of significant work in the field, but maintains that none of the subsequent manuals can effectively serve as substitutes to Euclid's early teachings of elementary geometry.

Cambridge University Press has long been a pioneer in the reissuing of out-of-print titles from its own backlist, producing digital reprints of books that are still sought after by scholars and students but could not be reprinted economically using traditional technology. The Cambridge Library Collection extends this activity to a wider range of books which are still of importance to researchers and professionals, either for the source material they contain, or as landmarks in the history of their academic discipline.

Drawing from the world-renowned collections in the Cambridge University Library, and guided by the advice of experts in each subject area, Cambridge University Press is using state-of-the-art scanning machines in its own Printing House to capture the content of each book selected for inclusion. The files are processed to give a consistently clear, crisp image, and the books finished to the high quality standard for which the Press is recognised around the world. The latest print-on-demand technology ensures that the books will remain available indefinitely, and that orders for single or multiple copies can quickly be supplied.

The Cambridge Library Collection will bring back to life books of enduring scholarly value (including out-of-copyright works originally issued by other publishers) across a wide range of disciplines in the humanities and social sciences and in science and technology.

Euclid and His Modern Rivals

CHARLES L. DODGSON

CAMBRIDGE
UNIVERSITY PRESS

CAMBRIDGE UNIVERSITY PRESS

Cambridge, New York, Melbourne, Madrid, Cape Town, Singapore,
São Paolo, Delhi, Dubai, Tokyo

Published in the United States of America by Cambridge University Press, New York

www.cambridge.org
Information on this title: www.cambridge.org/9781108001007

© in this compilation Cambridge University Press 2009

This edition first published 1879
This digitally printed version 2009

ISBN 978-1-108-00100-7 Paperback

EUCLID

AND HIS

MODERN RIVALS.

THE THEOREMS

OF

EUCLID I

Arranged in their Logical Sequence.

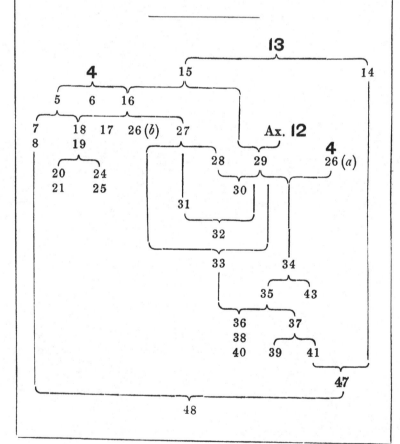

EUCLID

AND HIS

MODERN RIVALS

CHARLES L. DODGSON, M.A.

Senior Student and Mathematical Lecturer
of Christ Church, Oxford

'All for your delight
We are not here. *That you should here repent you*
The actors are at hand ; and, by their show,
You shall know all, that you are like to know.'

London

MACMILLAN AND CO.

1879

Dedicated
to
the memory
of
Euclid

PROLOGUE.

'ridentem dicere verum
Quid vetat?'

THE object of this little book is to furnish evidence, first, that it is essential, for the purpose of teaching or examining in elementary Geometry, to employ one text-book only; secondly, that there are strong *a priori* reasons for retaining, in all its main features, and specially in its sequence and numbering of propositions and in its treatment of parallels, the Manual of Euclid; and thirdly, that no sufficient reasons have yet been shown for abandoning it in favour of any one of the modern Manuals which have been offered as substitutes.

It is presented in a dramatic form, partly because it seemed a better way of exhibiting in alternation the arguments on the two sides of the question; partly that I might feel myself at liberty to treat it in a rather lighter style than would have suited an essay, and thus to make it a little less tedious and a little more acceptable to unscientific readers.

In one respect this book is an experiment, and may chance to prove a failure: I mean that I have not thought it necessary to maintain throughout the gravity of style which scientific writers usually affect, and which has somehow come to be regarded as an 'inseparable accident' of scientific teaching. I never could quite see the reasonableness of this immemorial law: subjects there are, no doubt, which are in their essence too serious to admit of any lightness of treatment—but I cannot recognise Geometry as one of them. Nevertheless it will, I trust, be found that I have permitted myself a glimpse of the comic side of things only at fitting seasons, when the tired reader might well crave a moment's breathing-space, and not on any occasion where it could endanger the continuity of a line of argument.

Pitying friends have warned me of the fate upon which I am rushing: they have predicted that, in thus abandoning the dignity of a scientific writer, I shall alienate the sympathies of all true scientific readers, who will regard the book as a mere *jeu d'esprit*, and will not trouble themselves to look for any serious argument in it. But it must be borne in mind that, if there is a Scylla before me, there is also a Charybdis—and that, in my fear of being read as a jest, I may incur the darker destiny of not being read at all.

In furtherance of the great cause which I have at heart —the vindication of Euclid's masterpiece—I am content to run some risk; thinking it far better that the purchaser of this little book should *read* it, though it be with a smile, than that, with the deepest conviction of its seriousness of purpose, he should leave it unopened on the shelf.

To all the authors, who are here reviewed, I beg to tender my sincerest apologies, if I shall be found to have transgressed, in any instance, the limits of fair criticism. To Mr. Wilson especially such apology is due—partly because I have criticised his book at great length and with no sparing hand—partly because it may well be deemed an impertinence in one, whose line of study has been chiefly in the lower branches of Mathematics, to dare to pronounce any opinion at all on the work of a Senior Wrangler. Nor should I thus dare, if it entailed my following him up ' yonder mountain height' which *he* has scaled, but which *I* can only gaze at from a distance : it is only when he ceases ' to move so near the heavens,' and comes down into the lower regions of Elementary Geometry, which I have been teaching for nearly five-and-twenty years, that I feel sufficiently familiar with the matter in hand to venture to speak.

Let me take this opportunity of expressing my gratitude, first to Mr. Todhunter, for allowing me to quote *ad libitum* from the very interesting Essay on Elementary Geometry, which is included in his volume entitled ' The Conflict of Studies, and other Essays on subjects connected with Education,' and also to reproduce some of the beautiful diagrams from his edition of Euclid ; secondly, to the Editor of the Athenæum, for giving me a similar permission with regard to a review of Mr. Wilson's Geometry, written by the late Professor De Morgan, which appeared in that journal, July 18, 1868.

ARGUMENT OF DRAMA.

ACT I.

Preliminaries to examination of Modern Rivals.

Scene I.

[Minos *and* Rhadamanthus.]

SCENE II.

[MINOS *and* EUCLID.]

§ 1. A priori *reasons for retaining Euclid's Manual.*

§ 3. *The combination, or separation, of Problems
and Theorems.*

§ 4. *Syllabus of propositions relating to Pairs of Lines.*

§ 5. *Playfair's Axiom.*

b

ACT II.

[MINOS *and* NIEMAND.]

*Manuals which reject Euclid's treatment
of Parallels.*

SCENE I.

b 2

SCENE IV.

Treatment of Parallels by equidistances.

CUTHBERTSON.

SCENE V.

Treatment of Parallels by direction.

§ 1. WILSON.

§ 2. Pierce.

§ 3. Willock.

ACT III.

Manuals which adopt Euclid's treatment of Parallels.

Scene I.

SCENE II.

§ 1. *Syllabus of the Association for the Improvement of Geometrical Teaching.*

§ 2. *Wilson's 'Syllabus'-Manual.*

     Of 73 propositions of Euclid, this Manual has
       14 omitted ;
       43 done as in Euclid ;
       10 done by new but objectionable methods, viz.—
         1 illogical ;
         1 'hypothetical construction' ;
         2 needlessly using 'superposition' ;
         2 algebraical ;
         4 omitting the diagonals of Euc. II. ;
       6 done by new and admissible methods.

ACT IV.

[MINOS *and* EUCLID.]

Manual of Euclid.

§ 1. *Treatment of Pairs of Lines.*

§ 2. *Euclid's constructions.*

§ 3. *Euclid's demonstrations.*

§ 4. *Euclid's style.*

§ 5. *Euclid's treatment of Lines and Angles.*

§ 6. *Omissions, alterations, and additions, suggested by Modern Rivals.*

§ 7. *The summing-up.*

APPENDICES.

I.

IV.

V.

VI.

ACT I.

SCENE I.

'Confusion worse confounded.'

[*Scene, a College study. Time, midnight.* MINOS *discovered seated between two gigantic piles of manuscripts. Ever and anon he takes a paper from one heap, reads it, makes an entry in a book, and with a weary sigh transfers it to the other heap. His hair, from much running of fingers through it, radiates in all directions, and surrounds his head like a halo of glory, or like the second Corollary of* Euc. I. 32. *Over one paper he ponders gloomily, and at length breaks out in a passionate soliloquy.*]

Min. So, my friend! *That's* the way you prove I. 19, is it? Assuming I. 20? Cool, refreshingly cool! But stop a bit! Perhaps he doesn't 'declare to win' on Euclid. Let's see. Ah, just so! 'Legendre,' of course! Well, I suppose I must give him full marks for it: what's the question worth?—Wait a bit, though! Where's his paper of yesterday? I've a very decided impression he was all for 'Euclid' then: and I know the paper had I. 20 in it.

B

... Ah, here it is! 'I think we do know the sweet Roman hand.' Here's the proposition, as large as life, and proved by I. 19. 'Now, infidel, I have thee on the hip!' You shall have such a sweet thing to do in *viva-voce*, my very dear friend! You shall have the two propositions together, and take them in any order you like. It's my profound conviction that you don't know how to prove either of them without the other. They'll have to introduce each other, like Messrs. Pyke and Pluck. But what fearful confusion the whole subject is getting into! (*Knocking heard.*) Come in!

Enter RHADAMANTHUS.

Rhad. I say! Are we bound to mark an answer that's a clear logical fallacy?

Min. Of course you are—with that peculiar mark which cricketers call 'a duck's egg,' and thermometers 'zero.'

Rhad. Well, just listen to this proof of I. 29.

Reads.

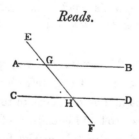

'Let *EF* meet the two parallel lines *AB*, *CD*, in the points *GH*. The alternate angles *AGH*, *GHD*, shall be equal.

'For *AGH* and *EGB* are equal because vertically opposite, and *EGB* is also equal to *GHD* (Definition 9); therefore *AGH* is equal to *GHD*, but these are alternate
angles.'

Did you ever hear anything like that for calm assumption?

Min. What does the miscreant mean by 'Definition 9'?

Rhad. Oh, that's the grandest of all! You must listen
to that bit too. There's a reference at the foot of the page
to 'Cooley.' So I hunted up Mr. Cooley among the heaps
of Geometries they've sent me—(by the way, I wonder if
they've sent *you* the full lot? Forty-five were left in my
rooms to-day, and ten of them I'd never even heard of
till to-day!)—well, as I was saying, I looked up Cooley,
and here's the Definition.

Reads.

'Right lines are said to be parallel when they are
equally and similarly inclined to the same right line, or
make equal angles with it towards the same side.'

Min. That is very soothing. So far as I can make
it out, Mr. Cooley quietly assumes that lines, which make
equal angles with *one* line, do so with *all*. If we had
Mr. Cooley in the Schools, I *think* we should pluck him.

Rhad. But as to this answer?

Min. Oh, give it full marks! What have we to do
with logic, or truth, or falsehood, or right, or wrong?
'We are but markers of a larger growth'—only that *we*
have to mark foul strokes, which a respectable billiard-
marker doesn't do as a general rule!

Rhad. There's one thing more I want you to look at.
Here's a man who puts 'Wilson' at the top of his paper,
and proves Euc. I. 32 from first principles, it seems to
me, without using any other theorem at all.

Min. The thing sounds impossible.

Rhad. So *I* should have said.   Here's the proof.

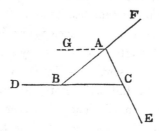

'Slide ∠ *DBA* along *BF* into position *GAF*, *GA* having
same direction as *DC* (Ax. 9); similarly slide ∠ *BCE*
along *AE* into position *GAC*.   Then the ext. ∠s = *CAF*,
FAG, *GAC* = one revolution = two straight ∠s.   But the
ext. and int. ∠s = 3 straight ∠s.   Therefore the int. ∠s =
one straight ∠ = 2 right angles.   Q. E. D.'

I'm not well up in 'Wilson': but surely he doesn't beg
the whole question of parallels in one axiom like this!

Min. Well, no.   There's a theorem and a corollary.   But
this is a sharp man: he has seen that the axiom does just
as well by itself.   Did you ever see one of those conjurers
bring a globe of live fish out of a pocket-handkerchief?
That's the kind of thing we have in Modern Geometry.
A man stands before you with nothing but an axiom in
his hands.   He rolls up his sleeves.   'Observe, gentlemen,
I have nothing concealed.   There is no deception!'   And

the next moment you have a complete theorem, Q. E. D. and all!

Rhad. Well, so far as *I* can see, the proof's worth nothing.   What am I to mark it?

Min. Full marks: we *must* accept it.   Why, my good fellow, I'm getting into that state of mind, I'm ready to mark *any* thing and *any* body.   If the Ghost in Hamlet came up this minute and said 'Mark me!' I should say 'I will!   Hand in your papers!'

Rhad. Ah, it's all very well to chaff, but it's enough to drive a man wild, to have to mark all this rubbish! Well, good night!   I must get back to my work.   [*Exit.*

Min. (*indistinctly*) I'll just take forty winks, and—

Snores.

ACT I.

Scene II.

Οὐκ ἀγαθὸν πολυκοιρανίη· εἷς κοίρανος ἔστω,
Εἷς βασιλεύς.

[Minos *sleeping: to him enter the Phantasm of* Euclid. Minos *opens his eyes and regards him with a blank and stony gaze, without betraying the slightest surprise or even interest.*]

---

§ 1. A priori *reasons for retaining Euclid's Manual.*

Euc. Now what is it you really require in a Manual of Geometry?

Min. Excuse me, but—with all respect to a shade whose name I have reverenced from early boyhood—is not that *rather* an abrupt way of starting a conversation? Remember, we are two thousand years apart in history, and consequently have never had a personal interview till now. Surely a few preliminary remarks—

Euc. Centuries are long, my good sir, but *my* time to-night is short: and I never was a man of many words. So kindly waive all ceremony and answer my question.

Min. Well, so far as I can answer a question that comes upon me so suddenly, I should say—a book that will exercise the learner in habits of clear definite conception, and enable him to test the logical value of a scientific argument.

Euc. You do *not* require, then, a complete repertory of Geometrical truth?

Min. Certainly not. It is the ἐνέργεια rather than the ἔργον that we need here.

Euc. And yet many of my Modern Rivals have thus attempted to improve upon me—by filling up what they took to be my omissions.

Min. I doubt if they are much nearer to completeness themselves.

Euc. I doubt it too. It is, I think, a friend of yours who has amused himself by tabulating the various theorems which might be enunciated in the single subject of Pairs of Lines. How many did he make them out to be?

Min. About two hundred and fifty, I believe.

Euc. At that rate, there would probably be, within the limits of my First Book, about how many?

Min. A thousand, at least.

Euc. What a popular school-book it will be! How boys will bless the name of the writer who first brings out the complete thousand!

Min. I think your Manual is fully long enough already for all possible purposes of teaching. It is not in the region of new matter that you need fear your Modern

Rivals: it is in *quality*, not in *quantity*, that they claim
to supersede you. Your methods of proof, so they assert,
are antiquated, and worthless as compared with the new
lights.

Euc. It is to that very point that I now propose to
address myself: and, as we are to discuss this matter
mainly with reference to the wants of *beginners*, we may
as well limit our discussion to the subject-matter of Books
I and II.

Min. I am quite of that opinion.

Euc. The first point to settle is whether, for purposes of
teaching and examining, you desire to have one fixed
logical sequence of propositions, or would allow the use
of conflicting sequences, so that one candidate in an examination might use X to prove Y, and another use Y to
prove X—or even that the same candidate might offer
both proofs, thus 'arguing in a circle.'

Min. A very eminent Modern Rival of yours, Mr. Wilson,
seems to think that no such fixed sequence is really necessary. He says (in his Preface, p. 10) 'Geometry when
treated as a science, treated inartificially, falls into a
certain order from which there can be no very wide
departure; and the manuals of Geometry will not differ
from one another nearly so widely as the manuals of
algebra or chemistry; yet it is not difficult to examine
in algebra and chemistry.'

Euc. Books may differ very 'widely' without differing
in logical sequence. Let me give you a few instances
of conflicting logical sequences in Geometry. Legendre
proves my Prop. 5 by Prop. 8, 18 by 19, 19 by 20, 27 by

28, 29 by 32. Cuthbertson proves 37 by 41. Reynolds proves 5 by 20. When Mr. Wilson has produced similarly conflicting sequences in the manuals of algebra or chemistry, we may then compare the subjects : till then, his remark is quite irrelevant to the question.

Min. I do not think he will be able to do so : indeed there are very few logical chains *at all* in those subjects— most of the propositions being proved from first principles. I think I may grant at once that it is essential to have *one* definite logical sequence, however many manuals we employ : to use the words of another of your Rivals, Mr. Cuthbertson (Pref. p. viii.), ' enormous inconvenience would arise in conducting examinations with no recognised sequence of propositions.' This however applies to *logical* sequences only, such as your Props. 13, 15, 16, 18, 19, 20, 21, which form a continuous chain. There are many propositions whose place in a manual would be partly arbitrary. Your Prop. 8, for instance, is not wanted till we come to Prop. 48, so that it might occupy any intermediate position, without involving risk of circular argument.

Euc. Now, in order to secure this uniform logical sequence, we should require to know, as to any particular proposition, what other propositions were its logical descendants, so that we might avoid using any of these in proving it ?

Min. Exactly so.

Euc. We might of course give this information by attaching to each enunciation references to its logical descendants : but this would be a very cumbrous plan.

A better way would be to give them in the form of
a genealogy, but this would be very bulky if the enunciations themselves were inserted : so that it would be desirable to have numbers to distinguish the enunciations.
In that case (supposing *my* logical sequence to be adopted)
the genealogy would stand thus :—(*see Frontispiece*).

Min. Would it not be enough to publish an arranged
list (which would be all the better if numbered also), and
to enact that no proposition should be used to prove any
of its predecessors ?

Euc. That would hamper the writers of manuals very
much more than the genealogy would. Suppose, for
instance, that you adopted, in the list, the order of
theorems in my First Book, and that a writer wished
to prove Prop. 8 by Prop. 47 : this would not interfere
with my logical sequence, and yet your list would bar him
from doing so.

Min. But we might place 8 close before 48, and he
would then be free to do as you suggest.

Euc. And suppose some other writer wished to prove
24 by 8 ?

Min. I see now that any single list must necessarily
prevent many possible arrangements which would not
conflict with the agreed-on logical sequence. And yet
this is what the Committee of the Association for the
Improvement of Geometrical Teaching have approved of,
namely, 'a standard sequence for examination purposes,'
and what the Association have published in their ' Syllabus of Plane Geometry.'

Euc. I think they have overlooked the fact that they

are enacting many more sequences, as binding on writers, than the one logical sequence which they desire to secure. Their list would be fitly supplemented by a 'standard' genealogy. But in any case we are agreed that it is desirable to have a standard list of enunciations, numbered for reference?

Min. We are.

Euc. The next point to settle is, *what* sequence and numbering to adopt. You will allow, I think, that there are strong *a priori* reasons for retaining *my* numbers. The propositions have been known by those numbers for two thousand years; they have been referred to probably by hundreds of writers—in many cases by the numbers only, without the enunciations : and some of them, I. 5 and I. 47 for instance—'the Asses' Bridge' and 'the Windmill'— are now historical characters, and their nicknames are 'familiar as household words.'

Min. Even if no better *sequence* than yours could be found, it might still be urged that a new set of *numbers* must be adopted, in order to introduce, in their proper places, some important theorems which have been added to the subject since your time.

Euc. That want, if it were proved to exist, might, I think, be easily provided for without discarding my system of numbers. If you wished, for instance, to insert two new propositions between I. 13 and I. 14, it would be far less inconvenient to call them 13 B and 13 C than to abandon the old numbers.

Min. I give up the objection.

Euc. You will allow then, I think, that my sequence

and system of numbers should not be abandoned without good cause?

Min. Oh, certainly. And the *onus probandi* lies clearly on your Modern Rivals, and not on you.

Euc. Unless, then, it should appear that one of my Modern Rivals, whose logical sequence is incompatible with mine, is so decidedly better in his treatment of really important topics, as to make it worth while to suffer all the inconvenience of a change of numbers, you would not recognise his demand to supersede my Manual?

Min. On that point let me again quote Mr. Wilson. In his Preface, p. 15, he says, 'In a few years I hope that our leading mathematicians will have published, perhaps in concert, one or more text-books of Geometry, not inferior, to say the least, to those of France, and that they will supersede Euclid by the sheer force of superior merit.'

Euc. And I should be quite content to be so superseded. 'A fair field and no favour' is all I ask.

§ 2. *Method of procedure in examining Modern Rivals.*

Min. You wish me then to compare your book with those of your Modern Rivals?

Euc. Yes. But, in doing this, I must beg you to bear in mind that a Modern Rival will not have proved his case if he only succeeds in showing

(1) that certain propositions might with advantage be omitted (for this a teacher would be free to do, so long as he left the logical sequence complete);

or (2) that certain proofs might with advantage be changed for others (for these might be interpolated as ' alternative proofs ') ;

or (3) that certain new propositions are desirable (for these also might be interpolated).

All these matters will need to be fully considered here-after, if you should decide that my Manual ought to be retained : but they do not constitute the evidence on which that decision should be based.

Min. That, I think, you have satisfactorily proved.   But what *would* you consider to be sufficient grounds for abandoning your Manual in favour of another ?

Euc. There are two defects, and two only, alleged to exist in my Manual, which I regard as *crucial* in this matter. The first concerns my arrangement of Problems and Theorems : the second my treatment of Parallels.

If it be agreed that Problems and Theorems ought to be treated separately, my system of numbers must of course be abandoned, and no reason will remain why my Manual should be retained as a *whole* ; which is the only point I am concerned with.   This question you can, of course, settle on its own merits, without examining any of the new Manuals.

If, again, it be agreed that, in treating Parallels, some other method, essentially different from mine, ought to be adopted, I feel that, after so vital a change as that, in-volving (as no doubt it would) the abandonment of my sequence and numbers, the remainder of my Manual would not be worth fighting for, though portions of it might be embodied in the new Manual. To settle this question, you must, of course, examine one by one the new methods that have been proposed.

Min. You would not even ask to have your Manual retained as an alternative for the new one?

Euc. No. For I think it essential for purposes of teaching, that in treating this vital topic one uniform method should be adopted; and that this method should be the best possible (for it is almost inconceivable that two methods of treating it should be *equally* good). An alternative proof of a minor proposition may fairly be inserted now and then as about equal in merit to the standard proof, and may make a desirable variety: but on this one vital point it seems essential that nothing but the best proof existing should be offered to the limited capacity of a learner. *Vacuis committere venis nil nisi lene decet.*

Min. I agree with you that we ought to have one system only, and that the best, for treating the subject of Parallels. But would you have me limit my examination of your 'Modern Rivals' to this single topic?

Euc. No. There are several other matters of so great importance, and admitting of so much variety of treatment, that it would be well to examine any method of dealing with them which differs much from mine—not with a view of substituting the new Manual for mine, but in order to make such changes in my proofs as may be thought desirable. There are other matters again, where changes have been suggested, which you ought to consider, but on *general* grounds, not by examining particular writers.

Let me enumerate what I conceive should be the subjects of your enquiry, arranged in order of importance.

(1) The combination, or separation, of Problems and Theorems.

(2) The treatment of Pairs of Lines, especially Parallels, for which various new methods have been suggested, viz. :—

(*a*) Proofs involving infinite series: suggested by LEGENDRE.

(β) Angles made with transversals : COOLEY.

(γ) Equidistance : CUTHBERTSON.

(δ) Direction : WILSON, PIERCE, WILLOCK.

(ε) The substitution of ' Playfair's Axiom ' for my Axiom 12.

If your decision, on these two crucial questions, be given in my favour, we may take it as settled, I think, that my Manual ought to be retained as a whole : how far it should be modified to suit modern requirements will be matter for further consideration.

(3) The principle of superposition.

(4) The use of diagonals in Book II.

These two are *general* questions, and will not need the examination of particular authors. After this, it will be well, in order that your enquiry into the claims of my

Modern Rivals may be as complete as possible, to review them one by one, with reference to their treatment of matters not already discussed, especially :—

    (5) Right Lines.
    (6) Angles, including right angles.
    (7) Propositions of mine omitted.
    (8) Propositions of mine treated by a new method.
    (9) New propositions.
    (10) And you may as well conclude, in each case, with a general survey of the book, as to style, &c.

The following may be taken as a fairly complete catalogue of the books to be examined :—

1. Legendre.	8. Loomis.
2. Cooley.	9. Morell.
3. Cuthbertson.	10. Reynolds.
4. Wilson.	11. Wright.
5. Pierce.	12. Wilson's ' Syllabus '-
6. Willock.	Manual.
7. Chauvenet.	

You should also examine the Syllabus, published by the Association for the Improvement of Geometrical Teaching, on which the last-named Manual is based. Not that it can be considered as a ' *Rival* '—in fact, it is not a text-book at all, but a mere list of enunciations—but because, first, it comes with an array of imposing names to recommend it, and secondly, it discards my system of numbers, so that its adoption, as a standard for examinations, would seriously interfere with the retention of my Manual as the standard text-book.

Now, of these questions, I shall be most happy to discuss with you the *general* ones (I mean questions 1, 2 (ε), 3, and 4) before we conclude this interview : but, when it comes to criticising particular authors, I must leave you to yourself, to deal with them as best you can.

Min. It will be weary work to do it all alone. And yet I suppose you cannot, even with *your* supernatural powers, fetch me the authors themselves?

Euc. I dare not. The living human race is so strangely prejudiced. There is nothing men object to so emphatically as being transferred by ghosts from place to place. I cannot say they are consistent in this matter : they are for ever 'raising' or 'laying' us poor ghosts—we cannot even haunt a garret without having the parish at our heels, bent on making us change our quarters : whereas if *I* were to venture to move one single small boy—say to lift him by the hair of his head over only two or three houses, and to set him down safe and sound in a neighbour's garden—why, I give you my word, it would be the talk of the town for the next month !

Min. I can well believe it. But what *can* you do for me? Are their *Doppelgänger* available?

Euc. I fear not. The best thing I can do is to send you the Phantasm of a German Professor, a great friend of mine. He has read all books, and is ready to defend any thesis, true or untrue.

Min. A charming companion! And his name?

Euc. Phantasms have no names—only numbers. You may call him 'Herr Niemand,' or, if you prefer it, 'Number one-hundred-and-twenty-three-million-four-hundred and-fifty-six-thousand-seven-hundred-and-eighty-nine.'

Min. For *constant* use, I prefer 'Herr Niemand.' Let us now consider the question of the separation of Problems and Theorems.

§ 3. *The combination, or separation, of Problems and Theorems.*

Euc. I shall be glad to hear, first, the reasons given for separating them, and will then tell you *my* reasons for mixing them.

Min. I understand that the Committee of the Association for the Improvement of Geometrical Teaching, in their Report on the Syllabus of the Association, consider the separation as 'equivalent to the assertion of the principle that, while Problems are from their very nature dependent for the form, and even the possibility, of their solution on the arbitrary limitation of the instruments allowed to be used, Theorems, being truths involving no arbitrary element, ought to be exhibited in a form and sequence independent of such limitations.' They add however that 'it is probable that most teachers would prefer to introduce Problems, not as a separate section of Geometry, but rather in connection with the Theorems with which they are essentially related.'

Euc. It seems rather a strange proposal, to print the propositions in one order and read them in another. But

a stronger objection to the proposal is that several of
the Problems are Theorems as well—such as I. 46, for
instance.

Min. How is that a Theorem?

Euc. It proves that there is such a thing as a square.
The definition, of course, does not assert real existence:
it is merely provisional. Now, if you omit I. 46, what
right have you, in I. 47, to say '*draw* a square'? How
do you know it to be possible?

Min. We could easily deduce that from I. 34.

Euc. No doubt a Theorem might be introduced for that
purpose: but it would be very like the Problem: you
would have to say '*if* a figure were drawn under such and
such conditions, it would be a square.' Is it not quite as
simple to draw it?

Then again take I. 31, where it is required to draw
a parallel. Although it has been proved in I. 27 that
such things as parallel Lines *exist*, that does not tell us
that, for every Line and for every point without that
Line, there exists a real Line, parallel to the given Line
and passing through the given point. And yet that is a fact
essential to the proof of I. 32.

Min. I must allow that I. 31 and I. 46 have a good
claim to be retained in their places: and if two are to
be retained, we may as well retain all.

Euc. Another argument, for retaining the Problems
where they are, is the importance of keeping the numbering unchanged—a matter we have already discussed.

But perhaps the strongest argument is that it saves
you from 'hypothetical constructions,' the danger of

which has been so clearly pointed out by Mr. Todhunter (see pp. 196, 205, 206, 217).

Min. I think you have proved your case very satisfactorily. The next subject is 'the treatment of Pairs of
Lines.' Would it not be well, before entering on this
enquiry, to tabulate the accepted truths respecting Pairs
of Lines?

Euc. That will be an excellent plan. It will both give
you a clear view of the field of your enquiry, and enable
you to recognise at once any doubtful axioms which you
may meet with.

Min. Will you then favour me with your views on this
matter?

Euc. Willingly. It is a subject which I need hardly say
I considered very carefully before deciding what definitions
and axioms to adopt.

§. 4. *Syllabus of propositions relating to Pairs of Lines.*

Let us begin with the simplest possible case, a Pair of
infinite Lines which have two common points, and which
therefore coincide wholly, and let us consider how such
a Pair may be defined, and what other properties it
possesses.

After that we will take a Pair of Lines which have
a common point and a separate point ('a separate point'
being one that lies on one of the Lines but not on the

other), and which therefore have no other common points, and treat it in the same way.

And in the third place we will take a Pair of Lines which have *no* common point.

Now the properties of a Pair of Lines may be ranged under four headings :—

(1) as to common or separate points ;

(2) as to the equality, or otherwise, of angles made with transversals ;

(3) as to the equidistance, or otherwise, of points on the one from the other ;

(4) as to direction.

We might distinguish the first two classes, which I have mentioned, as 'coincident' and 'intersecting' : and these names would serve very well if we were going to consider only infinite Lines; but, as all the relations of infinite Lines, with regard to angles made with transversals, equidistance of points, and direction, are equally true of finite portions of them, it will be well to use names which will include them also. And the names I would suggest are 'coincidental' and 'intersectional.'

By 'coincidental Lines,' then, I shall mean Lines which either coincide or would do so if produced : and by 'intersectional Lines' I shall mean Lines which either intersect or would do so if produced.

In the same way, when I speak of 'Lines having a common point,' or of 'Lines having two common points,' I shall mean Lines which either have such points or would have them if produced.

I will tabulate for you the propositions relating to Pairs of Lines, with the aid of some of those Algebraical symbols in which modern Mathematicians take so much delight.

Min. I should much like to see that done.

Euc. You will find the symbolical form much the most convenient for reference, but I will also give you the same tables printed in full, without symbols. (See App. III.) The symbols I would suggest are these :—

(1) As to points belonging to the Pair:—

Symbol.   *Meaning.*

c	one is common ;
cc	two are so ;
C	all are so ;
s	one is separate (i.e. the Lines are different);
S_{-1}	all but one are so (nothing asserted as to the excepted point);
S	all are so ;
X	one is common and the rest separate (i. e. it asserts ' cS_{-1}').

(2) As to angles made with the Pair by transversals :—

t_m	one makes equal angles ;
T_m	all do so ;
$t_{m'}$	one makes unequal angles ;
$T_{m'}$	all do so.

Particular properties under this heading may be distinguished by subscripts, as follows :—

Symbol.	*Meaning.*
	one transversal
t_a	makes equal alternate angles ;
t_e	makes equal exterior and interior opposite angles ;
t_i	makes two interior angles on the same side together equal to two right angles ;
t_r	is at right angles to both Lines.

[Observe that the subscripts are the initials of 'alternate,' 'exterior,' 'interior,' and 'right'; while the subscript '*m*,' which includes all these, is the first consonant of '*omnis.*']

$t_{a'}$	makes unequal alternate angles ;
$t_{e'}$	makes unequal exterior and interior opposite angles ;
$t_{i'}$	makes two interior angles on the same side together unequal to two right angles ;
$t_{r'}$	is at right angles to one Line, but not so to the other.

A double accent asserts what a single one does, but with additional details :—

$t_{a''}$	makes one of two alternate angles (viz. that which is exterior to the triangle formed by the Lines) greater than the other ;
$t_{e''}$	makes exterior angle of a triangle greater than interior opposite angle ;
$t_{i''}$	makes two interior angles of a triangle together less than two right angles.

(3) As to points lying on one of the Lines and on the same side of the other :—

Symbol.	Meaning.
e	two points on one Line, which lie on the same side of the other, are equidistant from it ;
E	all on each Line are so ;
f	two on one Line are not equidistant from the other ;
F	all on each Line, which lie on the same side of the other, are so ;
V	the Lines diverge without limit, i. e. a point may be found on each, whose distance from the other shall exceed any assigned length.

(4) As to the relative directions of the Lines :—

	they have a common point and
cD	coincidental directions ;
cD′	not coincidental directions.

[N.B. No meaning is, so far, assigned to ‘*D*’ or ‘*D′*’ when used of Lines *not* known to have a common point.]

Any accented letter denies what the letter asserts : e. g. ‘*c′*’ denies that the Lines have a common point, i.e. asserts ‘*S*.’ Also ℙ represents the word ‘proves.’

With regard to a set of three Lines, $\overline{a \mid \beta \mid \gamma}$ asserts that two of them have property ‘*a*,’ that another combination of two has property ‘*β*,’ and the remaining combination property ‘*γ*.’ As an example of the use of this notation, $\overline{a \mid \beta \mid}$ ℙ $\overline{\mid \mid \gamma}$ asserts that, if three Lines be such that two of them have property ‘*a*,’ and another two property 'β,' it may thence be proved that the remaining two have property 'γ.'

I will now arrange these symbols in two columns of ' contradictories ' :—

With these symbols we may work out the 'contranominal' of a proposition almost like a sum in Arithmetic. The simplest instance of a pair of contranominals is 'a proves β, not-β proves not-a'; which would be symbolically

c	S
cc	S_{-1}
C	s
t_m	$T_{m'}$
T_m	$t_{m'}$
e	F
E	f.

written '$a \,\mathbb{P}\, \beta, \beta' \,\mathbb{P}\, a'$'; where each property is affected with a 'not' and transposed to the opposite side. (This is curiously like the algebraical rule for transposing a quantity from one side of an equation to the other, with a change of sign.)

A more complicated instance would be '$a\beta\gamma \,\mathbb{P}\, \delta$,' where we may interchange δ with any one of the three, and so get three contranominals, '$\beta\gamma\delta' \,\mathbb{P}\, a', a\gamma\delta' \,\mathbb{P}\, \beta', a\beta\delta' \,\mathbb{P}\, \gamma'.$'

Min. Would '$a\delta' \,\mathbb{P}\, \beta'\gamma'$' be a legitimate contranominal?

Euc. No. $a\delta'$ proves that the *combination* $\beta\gamma$ is untrue : but this does not necessarily lead to $\beta'\gamma'$, which would imply *each* to be untrue. All we can say is that $a\delta'$ proves that some one of these three, $\beta'\gamma$, $\beta\gamma'$, $\beta'\gamma'$, is true; but we do not know which. The rule is, to transpose only *one* quantity at a time.

Min. We might then transpose $\beta\gamma$, considered as one *single* property ?

Euc. Certainly: and the resulting contranominal would be '$a\delta' \,\mathbb{P}\, (\beta\gamma)'$,' that is, 'all things, which are "a-and-not-δ," are *not* "β-and-γ."'

Min. What are the contranominals of ' a ℙ $\beta\gamma$ '?

Euc. That is merely one sentence containing two distinct propositions, ' a ℙ β ' and ' a ℙ γ.' Nothing is gained by writing them together.

Min. Will the same rules apply to the properties of three Lines?

Euc. Exactly the same: ' $\overline{a\,|\,\beta\,|}$ ℙ $\overline{|\ \ |\,\gamma}$ ' yields two contranominals, ' $\overline{a\,|\ \ |\,\gamma'}$ ℙ $\overline{|\,\beta'\,|\ }$,' and ' $\overline{\ \ |\,\beta\,|\,\gamma'}$ ℙ $\overline{a'\,|\ \ |\ }$.'

Min. We might get a third by taking a and β as a *single* assertion, might we not?

Euc. Yes. That would give us ' $\overline{\ \ |\ \ |\,\gamma'}$ ℙ $(\,\overline{a\,|\,\beta\,|}\ )'$.'

Min. Might not some contranominals fail of being true, owing to their subject being non-existent? For instance, if we know that ' $a\beta\gamma$ ℙ δ,' we may infer that ' $a\beta\delta'$ ℙ γ',' but this will be true only if $a\beta\delta'$ really exists.

Euc. A proposition, whose subject is unreal, is not necessarily *untrue*. It may be logically *valid*—that is, its predicate may be a necessary logical sequence of its subject. When we find that any contranominal proposition is unreal, though valid, it shows that *some one* of the data of the original proposition was superfluous.

Min. How do you make that out?

Euc. Why, we know that everything (and *a fortiori* every $a\beta$) must be one of the two, δ or δ': if then we are told that $a\beta\delta'$ is unreal, that is, that no $a\beta$ is δ', we may conclude that every one of them is δ, that is, that '$a\beta$ ℙ δ': so that γ was superfluous.

Min. I see that.

Euc. There is yet another fact which we may deduce

from '$a\beta\gamma$ ⚑ δ,' where $a\beta\gamma$ is real; namely, that $a\beta\gamma\delta$ is real also.

Now the propositions relating to Pairs of Lines may be divided into two classes, the first covering the ground occupied by my axiom 10 ('two straight lines cannot enclose a space') and my propositions I. 16, 17, 27, 28, 31; the second that occupied by my Axiom 12 and propositions I. 29, 30, 32. Those in the first class are logical deductions from axioms which have never been disputed: the second class has furnished, through all ages, a battle-field for rival mathematicians. That *some one* of the propositions in this class must be assumed as an axiom is agreed on all hands, and each combatant in turn proclaims his own special favourite to be the *one* axiomatic truth of the series, insisting that all the rest ought to be proved as theorems.

Let us now consider the properties of Pairs of Lines.

Such pairs may be arranged in three distinct classes. I will take them separately, and enumerate, for each class, first the 'subjects,' and secondly the 'predicates,' of propositions concerning it.

Min. Let us make sure that we understand each other as to those two words. I presume that a 'subject' will include just so much 'property' as is needed to indicate the Pair of Lines referred to, i.e. to serve as a sufficient definition for them?

Euc. Exactly so. Now, if we are told, with regard to a certain Pair of Lines, that

(1) they have two common points;

or (2) they have a common point, and make equal angles with a certain transversal;

or (3) they have a common point, and one of them
has two points on the same side of, and equidistant from,
the other;

or (4) they have a common point and coincidental
directions;

we may prove, with regard to this Pair of Lines, that
(1) they have all points common, i. e. they coincide
wholly;

and (2) they make equal angles with all transversals;

and (3) all points on each are equidistant from the other;

and (4) they have coincidental directions.

We assert, in short, that a Pair of Lines, which possesses any one of the properties '*cc*,' '*ct*$_m$,' '*ce*,' '*cD*,'
possesses all the properties '*C*,' '*T*$_m$,' '*E*,' '*D*.' Such a
Pair of Lines may be called 'coincidental.'

Min. There is something unsatisfactory in your second
'subject,' '*ct*$_m$.' Does it not include more than is necessary? I should have said that any one of these four
'*ct*$_a$,' '*ct*$_e$,' '*ct*$_i$,' '*ct*$_r$,' would have been a sufficient 'subject': but '*ct*$_m$' includes the whole four, does it not?

Euc. You are quite right. But you will admit that, if
any one of the four be given, it is a matter of the simplest
geometrical proof to deduce the other three?

Min. No doubt.

Euc. I think, then, that it will be convenient to agree
that, when '*t*$_m$' occurs in the *subject* of a proposition, it
shall mean '*any one* of the four t_a, t_e, t_i, t_r'; and that
when '*T*$_m$' occurs in the *predicate*, it shall mean '*all* the

four T_a, T_e, T_i, T_r.' Thus the proposition 't_m ℙ T_m' is equivalent to 't_a, or t_e, or t_i, or t_r, ℙ T_a and T_e and T_i and T_r'; so that it really involves *sixteen* propositions.

Min. That is clear enough.

Euc. There is, perhaps, a little obscurity in such a symbol as 't_i' when applied to a Pair of coincidental Lines: a diagram would make it clear. Let the lines AB, CD, in the first figure be supposed to coincide, and to become the Line KL in the second: then the two interior angles BGH, GHD, become the angles LMF, EML.

Min. Quite so.

Euc. Let us go on to the second class of Pairs of Lines.

If we are told, as to a certain Pair of Lines, that

(1) they have a common and a separate point;

or (2) they have a common point, and make unequal angles with a certain transversal;

or (3) they have a common point, and one of them has two points not-equidistant from the other;

or (4) they have a common point but not coincidental directions;

we may prove, with regard to this same Pair of Lines, that

(1) they have all other points separate;

and (2) they make unequal angles with all transversals;

and (3) every two points on each are not-equidistant from the other;

and (4) a point may be found on each, whose distance from the other shall exceed any assigned length;

and (5) they have not coincidental directions.

We assert, in short, that a Pair of Lines, which possesses any one of the properties 'cs,' '$ct_{m'}$,' 'cf,' 'cD',' possesses all the properties 'X,' '$T_{m'}$,' 'F,' 'V,' 'D'.' Such a Pair of Lines may be called 'intersectional.'

And thirdly, if we are told, with regard to a certain Pair of Lines, that

(1) they have a separate point, and make equal angles with a certain transversal;

or (2) they have a separate point, and one of them has two points on the same side of, and equidistant from, the other;

we may prove, with regard to this same pair of Lines, that they have no common point.

We assert, in short, that a Pair of Lines, which possesses either of the properties 'st_m,' 'se,' possesses the property 'S' (which might also be used as a definition). Such a Pair of Lines may be called 'separational.'

Min. Why not use your own word 'parallel'?

Euc. Because that word is not uniformly employed, by modern writers, in one and the same sense. Wilson, Willock, and Pierce, mean by it 'sD'; Cooley means either 'st_a' or 'sT_a,' it is not clear which ; and any writer on the 'equidistant' method might fairly adopt the meaning 'se' or 'sE.' I would advise you, in discussing the works of my Modern Rivals, to disallow the use of the word 'parallel' altogether, and to oblige each writer to adopt a word which shall express his own definition.

Min. There is one other question I wish to ask. In defining 'e' as asserting that 'two points on one Line, *which lie on the same side of the other*, are equidistant from it,' do you include the case of their lying *on* the other Line ?

Euc. Certainly. You may take them as lying on either side you like, and at zero-distances. The only case excluded is, where both points are *outside* the other Line, and on *opposite* sides of it.

Min. I understand you.

Euc. So far as we have now got, the subjects and predicates of propositions concerning the three Classes may be tabulated thus :—

	SUBJECTS.	PREDICATES.
I. Coincidental	ce, ct_m, ce, cD	C, T_m, E, D
II. Intersectional	cs, $ct_{m'}$, cf, cD'	X, $T_{m'}$, F, V, D'
III. Separational	st_m, se	S

The following Table contains nineteen propositions, of which some are axiomatic, some need to be proved but require only 'undisputed' axioms. I have placed contranominal propositions in the same Line :—

TABLE I. [See p. 227.]

Containing undisputed Axioms, and Theorems admitted to be real, valid, and deducible from undisputed Axioms.

COINCIDENTAL.	INTERSECTIONAL.	SEPARATIONAL.
1. $cc \, ℘ \, C$	2. (1) $s \, ℘ \, (cc)'$ or S_{-1} $\equiv cc$ unique (2) $cs \, ℘ \, X$	
3. $ct_m \, ℘ \, C$	4. $X \, ℘ \, T_{m'}$	5. $st_m \, ℘ \, S$
6. $ce \, ℘ \, C$	7. $X \, ℘ \, F$	8. $se \, ℘ \, S$
		9. sS real
	10. $X \, ℘ \, V$	
11. $cD \, ℘ \, C$	12. $X \, ℘ \, D'$	
	13. sD' real	
14. $cc \, ℘ \, D$	15. (1) $D' \, ℘ \, S_{-1}$ (2) $cD' \, ℘ \, X$	
16. $\overline{C|c|} \, ℘ \, \overline{||c}$	17. $\overline{c|c|S} \, ℘ \, \overline{X||}$ $\equiv \quad ℘ \, \overline{X|X|}$	18. $\overline{|c|S} \, ℘ \, \overline{s||}$
	19. $\overline{sE|cV|} \, ℘ \, \overline{| \ |X}$	

N.B. Under 4 are included :—

$$X \, ℘ \, T_{a'} \equiv \text{ext.} \angle \text{ of } \Delta \neq \text{int. alt.} \angle$$
$$X \, ℘ \, T_{e'} \equiv \text{ext.} \angle \text{ of } \Delta \neq \text{int. opp.} \angle$$
$$X \, ℘ \, T_{i'} \equiv 2 \text{ int.} \angle s \text{ of } \Delta \neq 2 \llcorner s$$
$$X \, ℘ \, T_{a''} \equiv \text{ext.} \angle \text{ of } \Delta > \text{int. alt.} \angle$$
$$X \, ℘ \, T_{e''} \equiv \text{ext.} \angle \text{ of } \Delta > \text{int. opp.} \angle$$
$$X \, ℘ \, T_{i''} \equiv 2 \text{ int.} \angle s \text{ of } \Delta < 2 \llcorner s$$

No. 19 requires some explanation, since one of the three Lines, to which it refers, is not assumed to be a *straight* Line. The Theorem asserts that, if there be given a straight Line and a point not on it; and if through the point there be drawn a second Line (*not* assumed to be straight) such that all points on it are equidistant from the first Line ; and if a third (straight) Line be drawn, having a point common with the first Line, and diverging from it without limit ; this third Line will intersect the second.

Min. That is quite satisfactory. I was just going to remark that No. 19 could not be taken to be a *real* Theorem, until we have proved that a real pair of Lines *can* have the property '*sE*.'

Euc. You are quite right.

Min. In 4, 7, 10, and 12, is not the 'subject' too extensive ? '*cs*' would be enough : you do not require '*X*.'

Euc. No doubt : but as the symbols are shown by 2 (2) to be interchangeable, each involving the other, it is simpler, I think, after proving 2 (2), to drop '*cs*' altogether, and to use '*X*' only.

Min. I observe that you have omitted two contranominals deducible from 11 and 15 (2).

Euc. You mean '*sD* ⚑ *S*' and '*X'D'* ⚑ *S*.' My reason for so doing is that we do not yet know either subject to be *real*.

Min. That is quite satisfactory.

Euc. We will now go a little further into the subject of separational Lines, as to which Table I. has furnished us with only three propositions. There are, however, many

other propositions concerning them, which are fully admitted to be *true*, though no one of them has yet been
proved from undisputed axioms: and we shall find that
they are so related to one another that, if any *one* be
granted as an axiom, all the rest may be proved; but,
unless some one be so granted, none can be proved. Two
thousand years of controversy have not yet settled the
knotty question *which* of them, if any, can be taken as
axiomatic.

If we are told, with regard to a certain Pair of Lines,
that either—

(1) they have no common point (i. e. all points belonging to the Pair are 'separate');

or (2) they have a separate point, and make equal angles
with a certain transversal;

or (3) they have a separate point, and one of them contains two points equidistant from the other;

we may prove (though not without the help of *some*
disputed axiom) that

(1) they make equal angles with all transversals;

and (2) all points on each are equidistant from the other.

That is, if we are told that they have any one of the properties ' S, st_m, se,' we may prove that they have both the
properties ' T_m, E.' These propositions, with the addition
of my own I. 30, I. 32, and certain others, I will now
arrange in a tabular form, placing contranominals in the
same line.

TABLE II.   [See p. 232.]

Consisting of Theorems admitted to be real and valid, but not *deducible from undisputed Axioms. They require, for proof, at least* one *Axiom, viz. either one of themselves, or else the proposition ' sD is real.'*

SEPARATIONAL.	INTERSECTIONAL.
1. (1) $\angle s$, which have sides of one separational from sides of other, each from each, are equal	2. (1) $t_{m'}$ ℙ c *(2) $st_{m'}$ ℙ X
(2) S ℙ T_m	
3. sT_m real	
4. st_m ℙ T_m	5. $st_{m'}$ ℙ $T_{m'}$
6. S ℙ E	7.* sf ℙ X
8.* sE real	
9. se ℙ T_m	10. $st_{m'}$ ℙ F
11. st_m ℙ.E	12. sf ℙ $T_{m'}$
13. se ℙ E	14. sf ℙ F
15. (1) $\overline{\,\mid S\mid S\,}$ ℙ $\overline{X'\mid}\,\mid$	16.*(1) $\overline{X\mid}\;\mid$ ℙ $\overline{(\;\mid S\mid S}$)'
(2) $\overline{c\mid S\mid S}$ ℙ $\overline{C\mid}\,\mid$	(2) $X\mid S\mid\;\overline{}$ ℙ $\overline{\;\mid\mid X}$
*≡ sS unique	
(3) $\overline{s\mid S\mid S}$ ℙ· $\overline{S\mid}\;\mid$	

17.* A Line cannot recede from and then approach another; nor can one approach and then recede from another on the same side of it.

18. (1) ext. $\angle$ of Δ = 2 int. opp. $\angle$ s.

(2) 3 $\angle$ s of Δ = 2 ∟ s.

Min. Would you be so kind as to translate No. 15 (1) into English ?

Euc. 'Two Lines, each of which is separational from a third Line, are not intersectional.'

Min. Why not say 'are separational'? Then it would exactly represent your I. 30.

Euc. Because they *might* be coincidental. My proposition is not quite accurately expressed : it ought to have been enunciated thus:—'*Different* Lines, which are parallel to the same Line, are parallel to each other.' Its true equivalent is No. 15 (3).

Min. Would you now translate No. 15 (2)?

Euc. 'Two Lines, which have a common point, and of which each is separational from a third Line, are coincidental.' This is equivalent to 'If, through a point outside a given Line, a Line be drawn separational from the first, such a Line is unique, i.e. only one such Line can be drawn.'

Min. Would you now translate No. 16 (2)?

Euc. 'A Line, which is intersectional with one of two separational Lines, is intersectional with the other.' It is contranominal to No. 15 (1).

Min. Is not the 'predicate' too extensive ? The true contradictory of '*S*' is '*c*,' so that your conclusion ought not to assert more than that they have a common point.

Euc. You are quite right: but we may fairly assume as an axiom that $\overline{X|S|}$ $\mathbb{P}$ $\overline{\;|\;|s}$, that is, that a Line, which is intersectional with (and so has a point common with) one of two separational Lines, has a point separate from the other. And this '*s*,' combined with the '*c*' which you say is the logical conclusion, gives ‘ *cs*,’ which justifies *my* conclusion ‘ *X*.’

Min. Some of these propositions seem to be much less axiomatic than others.

Euc. No doubt of it. The following have been proposed as Axioms (I have marked them with asterisks in the Table) :—

2 (2). ‘ $st_{m'}$ ℙ X ’; i.e. ‘ *Different Lines, which make un-equal angles with a certain transversal, are intersectional.*’ This is *my* choice, or rather a particular case of it, viz. $st_{i''}$ ℙ X; ‘ *Different Lines, which make with a transversal two interior angles together less than two right angles, are intersectional.*’

7. ‘ *sf* ℙ X ’; i.e. ‘ *Different Lines, of which one contains two points unequally distant from the other, are intersec-tional.*’ This is T. Simpson’s Axiom.

8. ‘ *sE real* ’; i.e. ‘ *Through a point, outside a given Line, a Line can be drawn such that the Pair shall be equidistant from each other.*’ This is virtually Clavius’ Axiom, though he only asserts the possibility of making the new Line equidistant from the given Line.

16 (1). ‘ $\overline{X|}$ | ℙ ($\overline{|S|S})'$ ’; i.e. ‘ *Intersectional Lines cannot both be separational from the same Line.*’ This is Playfair’s Axiom. Some writers prefer to assume its contranominal, viz. 15 (2). ‘ *sS* unique ’; i.e. ‘ *Through a given point, outside a given Line, only one Line can be drawn separational from it.*’

17. ‘ *A Line cannot recede from and then approach another;*

nor can one approach and then recede from another on the same side of it.' This might be called 'the Ahmedabad Axiom,' as it first appeared in an Arabic MS. of my Manual, which was found there in 1817: it has been adopted by R. Simpson and by Cuthbertson.

Min. I suppose I must take it on trust that any one of these 18 is sufficient logical basis for the other 17 : I can hardly ask you to go through 306 demonstrations !

Euc. I can do it with 13. You will grant me that, when two propositions are contranominal, so that each can be proved from the other, I may select either of the two for my series of proofs, but need not include *both* ?

Min. Certainly.

Euc. First I will give you 4, 3, 11, 8, 17, 13, 9, 18, in a cyclical series like a circulating decimal, each being proved by its predecessor only, and the last being used to prove the first.   Hence, if any one of these eight be given as an axiom, the rest can be proved from it.

Secondly, I will give you 16 (1), 1 (2), 6, in a similar series.

Thirdly, I will shew that, when either series has been proved, a member of the other series can be deduced.

Min. That will come to the same thing as a cyclical series made of the eighteen.

Euc. Here are the proofs, which you can read afterwards at your leisure. (See Appendix V.)   I will now give you two Tables of propositions relating to direction, which you will find useful when reviewing the writers who have adopted that theory.

You understand, of course, that I mean, by a 'real' proposition, one whose subject is admitted to be real; by an 'ideal' proposition, one whose subject is *not* admitted to be real; and, by a 'valid' proposition, one whose predicate is admitted to be a necessary sequence of the subject, whether that subject be real or not. For instance, 'a circular square has four corners' is a 'valid,' but 'ideal,' proposition.

Min. I understand.

Euc. In Table III, which I am about to lay before you, the subjects of the propositions are all 'ideal': hence, in this Table, the symbol '⊢' must not be taken to mean 'proves,' but '*would* prove if the subject really existed.'

TABLE III. [See p. 236.]

Consisting of valid but ideal *propositions; some being undisputed Axioms, the rest Theorems deducible from undisputed Axioms: but all depending for their* reality *on that of 'sD.'*

SEPARATIONAL.	INTERSECTIONAL.
1. sD ℙ S	
2. (1) ∠s, which have the sides of the one 'sD'-related to those of the other, each to each, are equal	
(2) sD ℙ T_m	
3. sD ℙ E	
4. | C |sD ℙ S | |	5. (1) c| |sD ℙ | s |
	(2) c | c |sD ℙ | X |
	≡ ℙ $\overline{X\,\vert\,X}$
6. (1) |sD|sD ℙ D | |	
(2) c |sD|sD ℙ cD| |	
(3) s |sD|sD ℙ sD| |	
7. (1) |sD|sD ℙ X'| |	8. X| | ℙ (|sD|sD)'
(2) c |sD|sD ℙ C | | ≡ sD unique	
(3) s |sD|sD ℙ S | |	

TABLE IV.   [See p. 238.]

Consisting of real *Theorems,* not *admitted to be valid : their validity depending on the reality of* '*sD.*'

SEPARATIONAL.	INTERSECTIONAL.
1. sD real	
2. $S \; \mathsf{P} \; D$	3. (1) $D' \; \mathsf{P} \; c$
	(2) $sD' \; \mathsf{P} \; X$
4. (1) An $\angle$ may be trans-	
ferred, preserving its	
magnitude and the	
directions of its sides	
(2) sDT_m real	
5. $st_m \; \mathsf{P} \; D$	6. $sD' \; \mathsf{P} \; T_{m'}$
7. $se \; \mathsf{P} \; D$	8. $sD' \; \mathsf{P} \; F$

§ 5. *Playfair's Axiom.*

Euc. The next *general* question to be discussed is the proposed substitution of Playfair's Axiom for mine. The question arises near the end of the series of propositions by which I prove I. 32.   This series (omitting elementary axioms) consists of Props. 4, 13, 15, 16, 27, 28, Ax. 12, Prop. 29.

Min. Down to the end of I. 28 there is no disputed point. My first question must be why you mention Ax. 12 *after* Prop. 28? It usually stands at the beginning of the Book.

Euc. Because it is not axiomatic until Prop. 28 has been proved. What is an axiom at one stage of our knowledge is often anything but an axiom at an earlier stage. Some of my critics, who have vehemently denied this to be an axiom, seem to have altogether overlooked the fact that there is no occasion whatever to ask the reader to assent to it until just before Prop. 29, where it is first wanted.

Min. There can be no objection to our postponing the consideration of it until the first twenty-eight propositions have been proved. The great question is whether it is axiomatic *then*.

Euc. I am quite aware of that : and it is because this is not only the great question of the whole First Book, but also the crucial test by which my method, as compared with those of my ' Modern Rivals,' must stand or fall, that I entreat your patience in speaking of a matter which cannot possibly be dismissed in a few words.

Min. Pray speak at whatever length you think necessary to so vital a point.

Euc. Let me remark in the first place—it is a minor matter, but yet one that *must* come in somewhere, and I do not want to break the thread of my argument—that we need, in any complete geometrical treatise, *some* practical geometrical test by which we can prove that two given finite Lines will meet if produced.   My 12th Axiom serves this purpose—a secondary purpose it is true—and it is incumbent on any one, who proposes to do away with this axiom, to provide some sufficient substitute.

Min. I admit all that.

Euc. My next remark will be best explained with the help of a diagram.

Let *AB* and *CD* make, with *EF*, the two interior angles *BEF, EFD*, together less than two right angles.   Now if through *E* we draw the Line *GH* such that the angles *HEF, EFD* may be equal to two right angles, it is easy to show (by Prop. 28) that *GH* and *CD* are 'separational.'

Min. Certainly.

Euc. We see, then, that any Lines which have the property (let us call it ' *a* ') of making, with a certain transversal, two interior angles together less than two

right angles, have also the property (let us call it ' β ')
that one of them intersects a Line which is separational
from the other.

Min. I grant it.

Euc. Now suppose you decline to grant my 12th Axiom,
but are ready to grant Playfair's Axiom, that two intersectional Lines cannot both be separational from the same
Line : then you have in fact granted my Axiom. The
two are in the same boat, and must sink or swim together.

Min. Be good enough to prove that.

Euc. Lines, which have property '*a*,' have property
' β.' Lines, which have property ' β,' meet if produced ;
for, if not, there would be two Lines both separational
from the same Line, which is absurd. Hence Lines,
which have property '*a*,' meet if produced.

Min. I see now that those who grant Playfair's Axiom
have no right to object to yours : but they may still
prefer Playfair's as the best.

Euc. I will now give you some reasons for preferring
mine.

In the first place, Playfair's Axiom (or rather the contranominal of it which I have been using, that ' a Line which
intersects one of two separational Lines will also meet the
other ') does not tell us *which way* we are to expect the
Lines to meet. But this is a very important matter in
constructing a diagram.

Min. We might obviate that objection by re-wording it
thus :—' If a Line intersect one of two separational Lines,
that portion of it which falls between them will also meet
the other.'

Euc. We might: and therefore I lay little stress on *that* objection.

In the second place, it is surely a more difficult conception for a beginner to grasp. *I* put before him a Pair of Lines, a transversal, and two angles whose sum is less than two right angles—all clear · *positive* conceptions. Playfair requires him to realise a Pair of Lines which never meet, though produced to infinity—a *negative* conception, which does not convey to the mind any clear notion of the relative position of the Lines.

Min. I am inclined to grant that yours has a slight advantage in respect of clearness of conception.

Euc. In the third place, Playfair's Axiom asserts *more* than mine does: and all the additional assertion is superfluous, and a needless strain on the faith of the learner.

Min. I do not see that in the least.

Euc. It *is* rather an obscure point, but I think I can make it clear. We know that all Pairs of Lines which have property 'a,' have also property 'β'; but we do *not* know as yet (till we have proved I. 29) that all which have property 'β,' have also property 'a.'

Min. That is so.

Euc. Then, for anything we know to the contrary, class 'β' *may* be larger than class 'a.' Hence, if you assert anything of class 'β,' the logical effect is more extensive than if you assert it of class 'a': for you assert it, not only of that portion of class 'β' which is known to be included in class 'a,' but also of the unknown (but possibly existing) portion which is *not* so included.

Min. I see that now, and consider it a real and very
strong reason for preferring your axiom.

But so far you have only answered Playfair. What do
you say to the objection raised by Mr. Potts? 'A stronger
objection appears to be that the converse of it forms Euc. I.
17; for both the assumed axiom and its converse should be
so obvious as not to require formal demonstration.'

Euc. Why, I say that I deny the general law which
he lays down. (It is, of course, the *technical* converse that
he means, not the *logical* one. 'All X is Y' has for its
technical converse 'All Y is X'; for its logical, 'Some Y
is X.') Let him try his law on the axiom 'The whole is
greater than a part,' and its technical converse 'Anything
greater than a part of a magnitude is equal to the whole
magnitude'!

Min. I withdraw the objection.

§ 6. *The Principle of Superposition.*

Min. The next subject is the principle of 'superposition.'
You use it twice only (in Props. 4 and 8) in the First
Book: but the modern fancy is to use it on all possible
occasions. The Syllabus indicates (to use the words of
the Committee) 'the free use of this principle as desirable
in many cases where Euclid prefers to keep it out of
sight.'

Euc. Give me an instance of this modern method.

Min. It is proposed to prove I. 5 by taking up the

isosceles triangle, turning it over, and then laying it down again *upon itself*.

Euc. Surely that has too much of the Irish Bull about it, and reminds one a little too vividly of the man who walked down his own throat, to deserve a place in a strictly philosophical treatise?

Min. I suppose its defenders would say that it is conceived to leave a trace of itself behind, and that the reversed triangle is laid down upon the trace so left.

Euc. That is, in fact, the same thing as conceiving that there are *two* coincident triangles, and that one of them is taken up, turned over, and laid down upon the other. And what does their subsequent coincidence prove? Merely this: that the right-hand angle of the first is equal to the left-hand angle of the second, and *vice versâ*. To make the proof complete, it is necessary to point out that, owing to the original coincidence of the triangles, this same 'left-hand angle of the second' is *also* equal to the *left*-hand angle of the first: and then, and not till then, we may conclude that the base-angles of the first triangle are equal. This is the full argument, strictly drawn out. The Modern books on Geometry often attain their much-vaunted brevity by the dangerous process of omitting links in the chain; and some of the new proofs, which at first sight seem to be shorter than mine, are really longer when fully stated. In this particular case I think you will allow that I had good reason for not adopting the method of superposition?

Min. You had indeed.

Euc. Mind, I do not object to that proof, if appended

to mine as an *alternative*. It will do very well for more advanced students. But, for beginners, I think it much clearer to have two non-isosceles triangles to deal with.

Min. But your objection to laying a triangle down *upon itself* does not apply to such a case as I. 24.

Euc. It does not. Let us discuss that case also. The Moderns would, I suppose, take up the triangle *ABC*, and apply it to *DEF* so that *AB* should coincide with *DE* ?

Min. Yes.

Euc. Well, that would oblige you to say 'and join *C*, in its new position, to *E* and *F*.' The words 'in its new position' would be necessary, because you would now have *two* points in your diagram, both called '*C*.' And you would also be obliged to give the points *D* and *E* additional names, namely '*A*' and '*B*.' All which would be very confusing for a beginner. You will allow, I think, that I was right here in constructing a new triangle instead of transferring the old one?

Min. Cuthbertson evades that difficulty by re-naming the point *C*, and calling it ' *Q*.'

Euc. And do the points *A* and *B* take their names with them?

Min. No. They adopt the names '*D*' and '*E*.'

Euc. It is very like making a new triangle !

Min. It is indeed. I think you have quite disposed of the claims of ' superposition.' The only remaining subject for discuss on is the omission of the diagonals in Book II.

§ 7.   *The omission of diagonals in Euc. II.*

Euc. Let us test it on my II. 4.   We will go through *my* proof of it, and then the proof given by some writer who ignores the diagonal, supplying if necessary any of those gaps in argument which my Modern Rivals so often indulge in, and which give to their proofs a delusive air of neatness and brevity.

'*If a line be divided into any two parts, the square of the line is equal to the squares of the two parts together with twice their rectangle.*

Let *AB* be a line, divided at *C*; then shall the square of *AB* be equal to the squares of *AC*, *CB*, with twice the rectangle of *AC*, *CB*.

On *AB* describe the square *ADEB*; join *BD*; from *C* draw *CF* parallel to *AD* or *BE*, cutting *BD* at *G*; and through *G* draw *HK* parallel to *AB* or *DE*.

∵ *BD* cuts the parallels *AD*, *CF*,

∴ the exterior angle *CGB*=the interior opposite angle *ADB*.   [I. 29

E

∵ $AD = AB$,

∴ the angle ADB=the angle ABD;   [I. 5

∴ the angle CGB=the angle ABD;

∴ $CG = CB$;   [I. 6

but $BK = CG$, and $GK = CB$;   [I. 34

∴ CK is equilateral.

It is also rectangular, since the angle CBK is right.

[I. 46. Cor.

∴ CK is a square.

Similarly HF is a square and =the square of AC. [I. 34

Also, ∵ AG, GE are equal, being complements,   [I. 43

∴ AG and GE=twice AG;

=twice the rectangle of AC, CB, since

$CG = CB$.

But these four figures make up AE, the square of AB.

Therefore the square of AB &c.   Q. E. D.'

That is just 136 words, counting from 'On AB describe' down to the words 'since $CG = CB$.' What author shall we turn to for a rival proof?

Min. I think Wilson will be best.

Euc. Very well. Do the best you can for him. You may use all my references if you like, and if you can do so legitimately.

Min. 'Describe a square $ADEB$ on AB. Through C draw CF parallel to AD, meeting——'

Euc. You must insert 'or BE,' to make the comparison fair.

Min. Certainly. I will mark the necessary insertions by parentheses. 'Through C draw CF parallel to AD (or BE), meeting DE in F.'

Euc. You may omit those four words, as they do not occur in my proof.

Min. Very well. 'Cut off (from CF) $CG=CB$. Through G draw HK parallel to AB (or DE). It is easily shewn that CK, HF are the squares of CB, AC; and that AG, GE, are each of them the rectangle of AC, CB.'

Euc. We can't admit ' it is easily shewn ' ! He is bound to *give* the proof.

Min. I will do it for him as briefly as I can. '$\because CG=CB$, and $BK=CG$, and $GK=CB$, $\therefore CK$ is equilateral. It is also rectangular, since the angle CBK is right. $\therefore CK$ is a square.' I'm afraid I mustn't say 'Similarly HF is a square'?

Euc. Certainly not: it requires a different proof.

Min. 'Because $CF=AD=AB$, and CG, CB, parts of them, are equal; $\therefore$ the remainder $GF=$the remainder $AC,=HG$. But $HD=GF$, and $DF=HG$; $\therefore HF$ is equilateral. It is also rectangular, since the angle HDF is right. $\therefore HF$ is a square, and$=$the square of AC. Also AG is the rectangle of AC, CB, since $CG=CB$.' I fear I can't assume GE to be equal to AG?

Euc. I fear I cannot permit you to assume the truth of my I. 43.

Min. 'Also GE is the rectangle of AC, CB, since $GF=AC$, and $GK=CB$. $\therefore AG$ and $GE=$twice the rectangle of AC, CB.'

Euc. That will do. How many words do you make it?

Min. 158.

Euc. Then the omission of the diagonal, instead of shortening the proof, has really lengthened it by twenty-two

words! Well! Has it any advantage in the way of neatness to atone for its greater length?

Min. Certainly not. It is quite unsymmetrical. I very
much prefer your method of appealing to the beautiful
theorem of the equality of complements.

Euc. Then that concludes our present interview: we will
meet again when you have reviewed my Modern Rivals
one by one. If you had any slow music handy, I would
vanish to it: as it is——

Vanishes without slow music.

ACT II.

Scene I.

'E fumo dare lucem.'

[Minos *sleeping. To him enter, first a cloud of tobacco-smoke ; secondly the bowl, and thirdly the stem, of a gigantic meerschaum ; fourthly the phantasm of* Herr Niemand, *carrying a pile of phantom-books, the works of Euclid's Modern Rivals, phantastically bound.*]

Niemand. The first author we have to consider is M. Legendre, is it not?

Minos. (*aside*) Not a single word of greeting! He plunges *in medias res* with a more fearful suddenness than Euclid himself! (*Aloud*) It is so, mein lieber Herr.

Nie. No time to waste in civil speeches! It is for you to question, for me to answer. I have read M. Legendre's book. Ach! It is beautiful! You shall find in it no flaw!

Min. I do not expect to do so.

ACT II.

Scene II.

Treatment of Parallels by methods involving infinite series.

Legendre.

'Fine by degrees, and beautifully less.'

Nie. I lay before you '*Éléments de Géométrie*' by Mons. A. M. Legendre, the 14th edition, 1860.

Min. Let me begin by asking you (since I consider you and your client as one in this matter) how you define a straight Line.

Nie. As 'the shortest path from one point to another.'

Min. This does not seem to me to embody the primary idea which the word 'straight' raises in the mind. Is not the natural process of thought to realise *first* the notion of 'a straight Line,' and *then* to grasp the fact that it is the shortest path between two points?

Nie. That may be the natural process : but surely you will allow our definition to be a legitimate one ?

Min. I think not : and I have the great authority of Kant to support me. In his ' Critique of Pure Reason,' he says (I quote from Meiklejohn's translation, in Bohn's Philosophical Library, pp. 9, 10), ' Mathematical judgments are always synthetical . . . " A straight line between two points is the shortest " is a synthetical proposition. For my conception of *straight* contains no notion of *quantity*, but is merely *qualitative*. The conception of the *shortest* is therefore wholly an addition, and by no analysis can it be extracted from our conception of a straight line.'

This may fairly be taken as a denial of the fitness of the axiom to stand as a definition. For all definitions ought to be the expressions of analytical, not of synthetical, judgments : their predicates ought not to introduce anything which is not already included in the idea corresponding to the subject. Thus, if the idea of 'shortest distance' cannot be obtained by a mere analysis of the conception represented by ' straight Line,' the axiom ought not to be used as a definition.

Nie. We are not particular as to whether it be taken as a definition or axiom : either will answer our purpose.

Min. Let us then at least banish it from the *definitions*. And now for its claim to be regarded as an *axiom*. It involves the assertion that a straight line is shorter than any *curved* line between the two points. Now the length of a curved line is altogether too difficult a subject for a beginner to have to consider : it is moreover unnecessary that he should consider it at all, at least in the earlier parts of Geometry : all he really needs is to grasp the fact that it is shorter than any *broken* line made up of straight lines.

Nie. That is true.

Min. And all cases of broken lines may be deduced from their simplest case, which is Euclid's I. 20.

Nie. Well, we will abate our claim and simply ask to have I. 20 granted us as an axiom.

Min. But it can be *proved* from your own axioms: and it is a generally admitted principle that, at least in dealing with beginners, we ought not to take as axiomatic any theorem 'which can be proved by the axioms we already possess.

Nie. For *beginners* we must admit that Euclid's method of treating this point is the best. But you will allow ours to be a legitimate and elegant method for the advanced student ?

Min. Most certainly. The whole of your beautiful treatise is admirably fitted for advanced students : it is only from the *beginner's* point of view that I venture to criticise it at all.

Your treatment of angles and right angles does not, I think, differ much from Euclid's ?

Nie. Not much. We *prove*, instead of assuming, that all right angles are equal, deducing it from the axiom that two right lines cannot enclose a space.

Min. I think some such proof a desirable interpolation.

I will now ask you how you prove Euc. I. 29 and 32.

Nie. What preliminary propositions will you grant us as proved ?

Min. Euclid's series consists of Props. 4, 13, 15, 16, 27, 28, Ax. 12, and Props. 29, 32. I will grant you as much of that series as you have proved by methods not radically differing from his.

Nie. That is, you grant us Props. 4, 13, and 15. Prop. 16 is not in our treatise. The next we require is Prop. 6.

Min. That you may take as proved.

Nie. And, next to that, Prop. 20 : *that* we assume as an axiom, and from it, with the help of Prop. 6, we deduce Prop. 19.

Min. For our present purpose you may take Prop. 19 as proved.

Nie. From Props. 13 and 19 we deduce Prop. 32 ; and, from that, Ax. 12 ; from which Prop. 29 follows at once.

Min. Your proof of Prop. 32 is long, but beautiful. I need not, however, enter on a discussion of its merits. It is enough to say that what we require is a proof suited to the capacities of *beginners*, and that this theorem of yours (Prop. XIX, at p. 20) contains an infinite series of triangles, an infinite series of angles, the terms of which continually decrease so as to be ultimately less than any assigned angle, and magnitudes which vanish simultaneously. These considerations seem to me to settle the question. I fear that your proof of this theorem, though a model of elegance and perspicuity as a study for the advanced student, is wholly unsuited to the requirements of a beginner.

Nie. That we are not prepared to dispute.

Min. It seems superfluous, after saying this, to ask what test for the meeting of Lines you have provided :

but we may as well have that stated, to complete the enquiry.

Nie. We give Euclid's 12th axiom, which we prove from Prop. 32, using the principle of Euc. X. 1 (second part), that 'if the greater of two unequal magnitudes be bisected, and if its half be bisected, and so on; a magnitude will at length be reached less than the lesser of the two magnitudes.'

Min. That again is a mode of proof entirely unsuited to beginners.

I have observed only one instance of what may be called faulty logic. It occurs in Prob. 1 (p. 49), where, in order to bisect a given Line, we are told to assume a length 'greater than half of it.' This would appear to require the previous solution of the Problem, and therefore is, strictly speaking, a '*Petitio Principii.*'

The general style of your admirable treatise I shall not attempt to discuss: it is one I would far rather take as a model to imitate than as a subject to criticise.

I can only repeat, in conclusion, what I have already said, that your book, though well suited for advanced students, is not so for beginners.

Nie. At this rate we shall make short work of the twelve Modern Rivals!

ACT II.

Scene III.

---

*Treatment of Parallels by angles
made with transversals.*

Cooley.

'The verbal solemnity of a hollow logic.'
Cooley, *Pref.* p. 20.

Nie. I have now the honour to lay before you '*The
Elements of Geometry, simplified and explained*,' by W. D.
Cooley, A.B., published in 1860.

Min. Please to hand me the book for a moment. I wish
to read you a few passages from the Preface. It is always
satisfactory—is it not?—to know that a writer, who
attempts to 'simplify' Euclid, begins his task in a be-
coming spirit of humility, and with some reverence for
a name that the world has accepted as an authority for
two thousand years.

Nie. Truly.

MINOS *reads.*

'The Elements of Plane Geometry . . . are here presented in the reduced compass of 36 propositions, perfectly
coherent, fully demonstrated, and reaching quite as far as
the 173 propositions contained in the first six books of
Euclid.' Modest, is it not?

Nie. A little high-flown, perhaps. Still, you know, if
they really *are* 'fully demonstrated'——

Min. If! In page 4 of the Preface he talks of 'Euclid's
circumlocutory shifts': in the same page he tells us that
'the doctrine of proportion, as propounded by Euclid, runs
into prolixity though wanting in clearness': and again, in
the same page, he states that most of Euclid's *ex absurdo*
proofs 'though containing little,' yet 'generally puzzle
the young student, who can hardly comprehend why
gratuitous absurdities should be so formally and solemnly
dealt with. These propositions therefore are omitted from
our Book of Elements, and the Problems also, for the
science of Geometry lies wholly in the Theorems. Thus
simplified and freed from obstructions, the truths of
Geometry may, it is hoped, be easily learned, even by
the youngest.' But perhaps the grandest sentence is at
the end of the Preface. 'Then as to those Propositions
(the first and last of the 6th Book), in which, according
to the same authority' (he is alluding to the Manual of
Euclid by Galbraith and Haughton), 'Euclid so beautifully
illustrates his celebrated definition, they appear to our
eyes to exhibit only the verbal solemnity of a hollow logic,
and to exemplify nothing but the formal application of

a nugatory principle.' Now let us see, mein Herr, whether
Mr. Cooley has done anything worthy of the writer of
such 'brave 'orts' (as Shakespeare has it): and first let
me ask how you define Parallel Lines.

NIEMAND reads.

'Right Lines are said to be parallel when they are
equally and similarly inclined to the same right Line, or
make equal angles with it towards the same side.'

Min. That is to say, if we see a Pair of Lines cut by
a certain transversal, and are told that they make equal
angles with it, we say 'these Lines are parallel'; and
conversely, if we are told that a Pair of Lines are parallel,
we say 'then there *is* a transversal, *somewhere*, which
makes equal angles with them'?

Nie. Surely, surely.

Min. But we have no means of finding it? We have
no right to draw a transversal at random and say '*this* is
the one which makes equal angles with the Pair'?

Nie. Ahem! Ahem! Ahem!

Min. You seem to have a bad cough.

Nie. Let us go to the next subject.

Min. Not till you have answered my question. *Have*
we any means of finding the particular transversal which
makes the equal angles?

Nie. I am sorry for my client, but, since you are so
exigeant, I fear I must confess that we have *no* means of
finding it.

Min. Now for your proof of Euc. I. 32.

Nie. You will allow us a preliminary theorem?

Min. As many as you like.

Nic. Well, here is our Theorem II. ' *When two parallel straight lines AB, CD, are cut by a third straight line EF, they make with it the alternate angles AGH, GHD, equal; and also the two internal angles at the same side BGH, GHD equal to two right angles.*

For *AGH* and *EGB* are equal because vertically opposite, and *EGB* is also equal to *GHD* (Definition); therefore—'

Min. There I must interrupt you. How do you know that *EGB* is equal to *GHD*? I grant you that, by the Definition, *AB* and *CD* make equal angles with *a certain* transversal: but have you any ground for saying that *EF is* the transversal in question?

Nic. We have not. We surrender at discretion. You will permit us to march out with the honours of war?

Min. We grant it you of our royal grace. March him off the table, and bring on the next Rival.

ACT II.

SCENE IV.

Treatment of Parallels by equidistances.

CUTHBERTSON.

'Thou art so near, and yet so far.'
Modern Song.

Nie. I now lay before you '*Euclidian Geometry,*' by
FRANCIS CUTHBERTSON, M.A., late Fellow of C. C. C. Cam-
bridge; Head Mathematical Master of the City of London
School; published in 1874.

Min. It will not be necessary to discuss with you *all* the
innovations of Mr. Cuthbertson's book. The questions of
the separation of Problems and Theorems, the use of superposition, and the omission of the diagonals in Book II, are
general questions which I have considered by themselves.
The only points, which you and I need consider, are the
methods adopted in treating Right Lines, Angles, and
Parallels, wherever those methods differ from Euclid's.

The first subject, then, is the Right Line. How do you
define and test it?

Nie. As in Euclid. But we *prove* what Euclid has assumed as an axiom, namely, that two right Lines cannot have a common segment.

Min. I am glad to hear you assert that Euclid has assumed it 'as an axiom,' for the interpolated and illogical corollary to Euc. I. 11 has caused many to overlook the fact that he has assumed it as early as Prop. 4, if not in Prop. 1. What is your proof?

Niemand reads.

' *Two straight lines cannot have a common segment.*'

' For if two straight lines *ABC*, *ABH* could have a common segment *AB* ; then the straight line *ABC* might be turned about its extremity *A*, towards the side on which *BH* is, so as to cut *BH* ; and thus two straight lines would enclose a space, which is impossible.'

Min. You assume that, *before C* crosses *BH*, the portions coinciding along *AB* will diverge. But, if *ABH* is a right Line, this will not happen till *C* has *passed H*.

Nie. But you would then have one portion of the revolving Line in motion, and another portion at rest.

Min. Well, why not?

Nie. We may assume *that* to be impossible ; and that,

if a Line revolves about its extremity, it all moves at once.

Min. Which, I take the liberty to think, is *quite* as great an assumption as Euclid's. I think the axiom quite plain enough without any proof.

Your treatment of angles, and right angles, is the same as Euclid's, I think ?

Nie. Yes, except that we *prove* that ' all right angles are equal.'

Min. Well, it *is* capable of proof, and therefore had better not be retained as an axiom.

I must now ask you to give me your proof of Euc. I. 32.

Nie. We prove as far as I. 28 as in Euclid. In order to prove I. 29, we first prove, as a Corollary to Euc. I. 20, that ' the shortest distance between two points is a straight line.'

Min. What is your next step ?

Nie. A problem (Pr. F. p. 52) in which we prove the theorem that, of all right lines drawn from a point to a Line, the perpendicular is the least.

Min. We will take that as proved.

Nie. We then deduce that the perpendicular is the shortest path from a point to a Line.

Next comes a Definition. ' By the distance of a point from a straight line is meant the shortest path from the point to the line.'

Min. Have you anywhere defined the distance of *one* point from another ?

Nie. No.

F

Min. We had better have that first.

Nie. Very well. 'The distance of one point from another is the shortest path from one to the other.'

Min. Might we not say ' is the length of the right line joining them ?'

Nie. Yes, that is the same thing.

Min. And similarly we may modify the definition you gave just now.

Nie. Certainly. 'The distance of a point from a right Line is the length of the perpendicular let fall upon it from the given point.'

Min. What is your next step ?

Niemand *reads.*

P. 33. *Ded. G.* 'If points be taken along one of the arms of an angle farther and farther from the vertex, their distances from the other arm will at length be greater than any given straight line.'

In proving this we assume as an axiom that the lesser of two magnitudes of the same kind can be multiplied so as to exceed the greater.

Min. I accept the axiom and the proof.

Niemand *reads.*

P. 34. *Ax.* 'If one right Line be drawn in the same plane as another, it cannot first recede from and then approach to the other, neither can it first approach to and then recede from the other on the same side of it.'

Min. Here, then, you assume, as axiomatic, one of the propositions of Table II. After this, you ought to have no further difficulty in proving Euc. I. 32 and all other properties of parallels. How do you proceed?

Nie. We prove (p. 34. *Lemma*) that, if two Lines have a common perpendicular, each is equidistant from the other.

Min. (*aside*) ' *st,* **P** *E.*' (*aloud*) What then?

Nie. Next, that any Line intersecting one of these will intersect the other (p. 35).

Min. That, I think, depends on Deduction *G*, at p. 33?

Nie. Yes.

Min. A short, but not very easy, Theorem; and one containing a somewhat intricate diagram. However, it proves the point. What is your next step?

NIEMAND *reads.*

P. 34. Lemma. 'Through a given point without a given straight line one and only one straight line can be drawn in the same plane with the former, which shall never meet it. Also all the points in each of these straight lines are equidistant from the other.'

Min. I accept all that.

Nie. We then introduce Euclid's definition of ' Parallels.' It is of course now obvious that parallel Lines are equidistant, and that equidistant Lines are parallel.

Min. Certainly.

Nie. We can now, with the help of Euc. I. 27, prove I. 29, and thence I. 32.

Min. No doubt. We see, then, that you propose, as a substitute for Euclid's 12th Axiom, a new Definition, two new Axioms, and what virtually amounts to five

new Theorems. In point of 'axiomaticity' I do not think
there is much to choose between the two methods. But
in point of brevity, clearness, and suitability to a beginner,
I give the preference altogether to Euclid's axiom.

The next subject to consider is your practical test, if
any, for two given Lines meeting when produced.

Nie. One test is that one of the lines should meet a
Line parallel to the other.

Min. Certainly : and that will suffice in such a case as
Euc. I. 44 (Pr. M. p. 60, in this book) though you omit
to point out *why* the Lines may be assumed to meet.
But what if the diagram does not contain 'a Line parallel
to the other'? Look at Pr. (*h*) p. 69, where we are told to
make, at the ends of a line, two angles which are together
less than two right angles, and where it is assumed that the
Lines, so drawn, will meet. That is, you assume the truth
of Euclid's 12th axiom. And you do the same thing at
pp. 70, 123, 143, and 185.

Nie. Euclid's 12th axiom is easily proved from our
theorems.

Min. No doubt: but you have not done it, and the omission makes a very serious hiatus in your argument. It is
not a thing that beginners are at all likely to be able to
supply for themselves.

I have no adverse criticisms to make on the general
style of the book, which seems clear and well written.
Nor is it necessary to discuss the claims of the book to
supersede Euclid, since the writer makes no such claim,
but has been careful (as he states in his preface) to avoid
any arrangement incompatible with Euclid's order. The

chief novelty in the book is the introduction of the principle of 'equidistance,' which does not seem to me a desirable feature in a book meant for beginners : otherwise it is little else than a modified version of Euclid.

ACT II.

Scene V.

Treatment of Parallels by direction.

---

§ 1. Wilson.

'There is moreover a logic besides that of mere reasoning.'
Wilson, *Pref.* p. xiii.

Nie. You have made but short work of four of the five methods of treating Parallels.

Min. We shall have all the more time to give to the somewhat intricate subject of Direction.

Nie. I lay on the table '*Elementary Geometry,*' by J. M. Wilson, M.A., late Fellow of St. John's College, Cambridge, late Mathematical Master of Rugby School, now Head Master of Clifton College. The second edition, 1869. And I warn you to be careful how you criticise it, as it is already adopted in several schools.

Min. Tant pis pour les écoles. So you and your client deliberately propose to supersede Euclid as a text-book?

Nie. 'I am of opinion that the time is come for making an effort to supplant Euclid in our schools and universities.' (Pref. p. xiv.)

Min. It will be necessary, considering how great a change you are advocating, to examine your book *very* minutely and critically.

Nie. With all my heart. I hope you will show, in your review, 'the spirit without the prejudices of a geometrician.' (Pref. p. xv.)

Min. We will begin with the Right Line. And first, let me ask, how do you define it?

Nie. As 'a Line which has the same direction at all parts of its length.' (p. 3.)

Min. You do not, I think, make any practical use of that as a test, any more than Euclid does of the property of lying evenly as to points on it?

Nie. No, we do not.

Min. You construct and test it as in Euclid, I believe? And you have his axiom that 'two straight Lines cannot enclose a space?'

Nie. Yes, but we extend it. Euclid asserts, in effect, that two Lines, which coincide in two points, coincide *between* those points : *we* say they 'coincide wholly,' which includes coincidence *beyond* those points.

Min. Euclid tacitly assumes that.

Nie. Yes, but he has not expressed it.

Min. I think the addition a good one. Have you any other axioms about it?

Nie. Yes, 'that a straight Line marks the shortest distance between any two of its points.' (p. 5. Ax. I.)

Min. That I have already fully discussed in reviewing
M. Legendre's book (see p. 54).

Nie. We have also 'A Line may be conceived as transferred from any position to any other position, its magnitude being unaltered.' (p. 5. Ax. 3.)

Min. True of *any* geometrical magnitude : but hardly
worth stating, I think. I have now to ask you how you
define an Angle?

Nie. 'Two straight Lines that meet one another form
an angle at the point where they meet.' (p. 5.)

Min. Do you mean that they form it '*at* the point' and
nowhere else?

Nie. I suppose so.

Min. I fear you allow your angle no magnitude, if you
limit its existence to so small a locality!

Nie. Well, we *don't* mean ' nowhere else.'

Min. (*meditatively*) You mean '*at* the point—and *some-*
where else.' *Where* else, if you please?

Nie. We mean—we don't quite know why we put in the
words at all. Let us say 'Two straight Lines that meet
one another form an angle.'

Min. Very well. It hardly tells us what an angle *is*,
and, so far, it is inferior to Euclid's definition : but it may
pass. Do you put any limit to the *size* of an angle?

Nie. We have not named any, but the largest here
treated of is what we call ' one revolution.'

Min. You admit reëntrant angles then?

Nie. Yes.

Min. Then your definition only states half the truth :
you should have said ' form *two* angles.'

Nie. That would be true, no doubt.

Min. But this extension of limit will require several modifications in Euclid's language: for instance, what is your definition of an obtuse angle?

NIEMAND *reads.*

P. 8. Def. 13. 'An *obtuse angle* is one which is greater than a right angle.'

Min. So you tumble headlong into the very first pitfall you come across! Why, that includes such angles as 180° and 360°. You would teach your pupil, I suppose, that one portion of a straight Line makes an obtuse angle with the other, and that every straight Line has an obtuse angle at each end of it!

Nie. It is an oversight—of course we ought to have added 'but less than two right angles.'

Min. A very palpable oversight. I fear we shall find more as we go on. What axioms have you about angles?

NIEMAND *reads.*

P. 5. Ax. 4. 'An angle may be conceived as transferred to any other position, its magnitude being unaltered.'

Min. Hardly worth stating. Proceed.

NIEMAND *reads.*

P. 5. Ax. 5. 'Angles are equal when they could be placed on one another so that their vertices would coincide in position, and their arms in direction.'

Min. 'Placed on one another'! Did you ever see the child's game, where a pile of four hands is made on the

table, and each player tries to have a hand at the top
of the pile ?

Nie. I know the game.

Min. Well, did you ever see both players succeed at
once ?

Nie. No.

Min. Whenever that feat is achieved, you may *then*
expect to be able to place two angles 'on one another'!
You have hardly, I think, grasped the physical fact that,
when one of two things is *on* the other, the second is
underneath the first. But perhaps I am hypercritical.
Let us try an example of your axiom : let us place an
angle of 90° on one of 270°. I think I could get
the vertices and arms to coincide in the way you
describe.

Nie. But the one angle would not be *on* the other ;
one would extend round one-fourth of the circle, and the
other round the remaining three-fourths.

Min. Then, after all, the angle is a mysterious entity,
which extends from one of the lines to the other? That
is much the same as Euclid's definition. Let us now take
your definition of a Right Angle.

Nie. We first define 'one revolution,' which is the angle
described by a line revolving, about one extremity, round
into its original position.

Min. That is clear enough.

Nie. We then say (p. 7. Def. 9) 'When it coincides with
what was initially its continuation, it has described *half
a revolution*, and the angle it has then described is called
a straight angle.'

Min. How do you know that it has described *half a revolution?*

Nie. Well, it is not difficult to prove. Let that portion of the plane, through which it has revolved, be rolled over, using as an axis the arm (in its initial position) and its continuation, until it falls upon the other portion of the plane. The two angular magnitudes will now together make up 'one revolution': therefore each is 'half a revolution.'

Min. A proof, I grant: but you are *very* sanguine if you expect beginners in the subject to supply it for themselves.

Nie. It is an omission, we admit.

Min. And then 'a straight angle'! 'Straight' is necessarily unbending: while 'angle' is from ἄγκος, 'a bend or hook': so that your phrase is exactly equivalent to 'an unbending bend'! In 'the Bairnslea Foaks' Almanack' I once read of 'a mad chap' who spent six weeks 'a-trying to maäk a straät hook': but he failed. He ought to have studied your book. Have you Euclid's axiom 'all right angles are equal'?

Nie. We deduce it from 'all straight angles are equal': and that we prove by applying one straight angle to another.

Min. That is all very well, though I cannot think 'straight angles' a valuable contribution to the subject. I will now ask you to state your method of treating Pairs of Lines, as far as your proof of Euc. I. 32.

Nie. To do that we shall of course require parallel Lines: and, as our definition of them is 'Lines having the same direction,' we must begin by discussing direction.

Min. Undoubtedly. How do you define direction?

Nie. Well, we have not attempted *that*. The idea seemed to us to be too elementary for definition. But let me read you what we have said about it.

Reads.

P. 2. Def. 2. ' A *geometrical Line* has position, and length, and at every point of it it has direction . . . .'

P. 3. Def. 4. ' A *straight Line* is a Line which has the *same* direction at all parts of its length. It has also the opposite direction . . . . A straight Line may be conceived as generated by a point moving always in the *same* direction.'

I will next quote what we have said about two Lines having ' the same direction ' and ' different directions.'

Min. We will take that presently : I have a good deal to say first as to what you have read. I gather that you consider direction to be a *property* of a geometrical entity, but not itself an entity ?

Nie. Just so.

Min. And you ascribe this property to a Line, and also to the motion of a point ?

Nie. We do.

Min. For simplicity's sake, we will omit all notice of curved Lines, etc., and will confine ourselves to straight Lines and rectilinear motion, so that in future, when I use the word ' Line,' I shall mean ' straight Line.' Now may we not give a notion of ' direction ' by saying—that a moving point must move in a certain 'direction'—that, if two points, starting from a state of coincidence, move along two equal straight Lines which do not coincide (so that their movements are alike in point of departure, and in magnitude), that quality of each movement, which makes it differ from the other, is its 'direction'—and similarly that, if two equal straight Lines are terminated at the same point, but do not coincide, that quality of each which makes it differ from the other, is its 'direction' from the common point?

Nie. It is all very true : but you are using 'straight Line' to help you in defining 'direction.' *We*, on the contrary, consider 'direction' as the more elementary idea of the two, and use it in defining 'straight Line.' But we clearly agree as to the meanings of both expressions.

Min. I am satisfied with that admission. Now as to the phrase 'the same direction,' which you have used in reference to a single Line and the motion of a single point. May we not say that portions of the same Line have 'the same direction' as one another? And that, if a point moves along a Line without turning back, its motion at one instant is in 'the same direction' as its motion at another instant?

Nie. Yes. That expresses our meaning in other language.

Min. I have altered the language in order to bring out clearly the fact that, in using the phrase 'the same direction' we are really contemplating *two* Lines, or *two* motions. We have now got (considering 'straight Line' as an understood phrase) accurate geometrical definitions of at least *two* uses of the phrase. And to these we may add a third, viz. that two coincident Lines have 'the same direction.'

Nie. Certainly, for they are one and the same Line.

Min. And you intend, I suppose, to use the word 'different' as equivalent to 'not-same.'

Nie. Yes.

Min. So that if we have, for instance, two equal Lines terminated at the same point, but not coinciding, we say that they have 'different directions'?

Nie. Yes, with one exception. If they are portions of one and the same infinite Line, we say that they have 'opposite directions.' Remember that we said, of a Line, ' it has also the opposite direction.'

Min. You did so: but, since 'same' and 'different' are contradictory epithets, they must together comprise the whole genus of 'pairs of directions.' Under which heading will you put 'opposite directions'?

Nie. No doubt, strictly speaking, 'opposite directions' are a particular kind of 'different directions.' But we shall have endless confusion if we include them in that class. We wish to avoid the use of the word 'opposite' altogether, and to mean, by 'different directions,' all kinds of directions that are not the same, with the exception of 'opposite.'

Min. It is a most desirable arrangement: but you have not clearly stated it in your book. Tell me whether you agree in this statement of the matter. Every Line has a pair of directions, opposite to each other. And if two Lines be said to have 'the same direction,' we must understand

' the same *pair* of directions ' ; and if they be said to have ' different directions,' we must understand ' different *pairs* of directions.'   And even this is not enough : for suppose I draw, on the map of England, a straight Line joining London and York ; I may say ' This Line has a pair of directions, the first being " London-to-York" and the second " York-to-London." '   I will now place another Line upon this, and *its* pair of directions shall be, first " York-to-London" and second " London-to-York."   Then it has a different first-direction from the former Line, and also a different second-direction : that is, it has a ' different pair of directions.'   Clearly *this* is not intended : but, in order to exclude such a possibility, we must extend yet further the meaning of the phrase, and, if two Lines be said to have ' the same direction,' we must understand ' pairs of directions which can be arranged so as to be the same ' ; and if they be said to have ' different directions,' we must understand ' pairs of directions which cannot be arranged so as to be the same.'

Nie. Yes, that expresses our meaning.

Min. You must admit, I think, that your theory of direction involves a good deal of obscurity at the very outset.   However, we have cleared it up, and will not use the word ' opposite ' again.   Tell me now whether you accept this as a correct definition of the phrases ' the same direction ' and ' different directions,' when used of a pair of infinite Lines which have a common point :—

If two infinite Lines, having a common point, coincide, they have ' the same direction ' ; if not, they have ' different directions.'

Nie. We accept it.

Min. And, since a finite Line has the same direction as the infinite Line of which it is a portion, we may generalise thus :—' Coincidental Lines have the same direction. Non-coincidental Lines, which have a common point, have different directions.' (Tab. I. 14, 12.)

But it must be carefully borne in mind that we have as yet no geometrical meaning for these phrases, *unless when applied to two Lines which have a common point.*

Nie. Allow me to remark that what *you* call ' coincidental Lines ' *we* call ' the same Line' or ' parts of the same Line,' and that what *you* call ' non-coincidental Lines ' *we* call ' different Lines.'

Min. I understand you : but I cannot employ these terms, for two reasons : first, that your phrase ' the same Line' loses sight of a fact I wish to keep in view, that we are considering a *pair* of Lines ; secondly, that your phrase ' different Lines' might be used, with strict truth, of two different portions of the same infinite Line, so that it is not definite enough for my purpose.

Let us now proceed ' to consider the relations of two or more straight Lines in one plane in respect of direction.'

And first let me ask which of the propositions of Table II you wish me to grant you as an axiom ?

Nie. (*proudly*) Not one of them! We have got a new patent process, the ' direction ' theory, which will dispense with them all.

Min. I am *very* curious to hear how you do it.

NIEMAND *reads.*

P. 11. Ax. 6. ' Two different Lines may have either the same or different directions.'

Min. That contains two assertions, which we will consider separately.   First, you say that ' two different Lines (i. e. ' non-coincidental Lines,' or ' Lines having a separate point') ' may have the same direction '?

Nie. Just so.

Min. That is, you assert ' *sD* is real.'   Now let us understand each other quite clearly.   We will take a fixed Line to begin with, and a certain point on it : there is no doubt that we can draw, through that point, a second Line coinciding with the first: the direction of this Line will of course be ' the same ' as the direction of the first Line ; and it is equally obvious that if we draw the second Line in any *other* direction, so as *not* to coincide with the first, its direction will *not* be ' the same ' as that of the first : that is, they will have ' different ' directions.   If we want a geometrical definition of the assertion that this second Line has ' the same direction ' as the first Line, we may take the following :—' having such a direction as will cause the Lines to be the same Line.'   If we want a geometrical construction for it, we may say ' take any other point on the fixed Line ; join the two points, and produce the Line, so drawn, at both ends' : this construction we know will produce a Line which will be ' the same ' as the first Line, and whose direction will therefore be ' the same ' as that of the first Line.   If, in a certain diagram, whose geometrical history we know, we want to test

G

whether two Lines, passing through a common point, have, or have not, 'the same direction,' we have simply to take any other point on one of the Lines, and observe whether the other Line does, or does not, pass through it. This relationship of direction, which *you* denote by the phrase 'having the same direction,' and *I* by the phrase 'having coincident directions,' we may symbolise by the letter '*D*.'

Nie. All very true. My only puzzle is, why you have explained it at such enormous length : my meerschaum has gone out while I have been listening to you!

Min. Allow me to hand you a light. As to the 'enormous length' of my explanation, we are in troubled waters, my friend! There are breakers ahead, and we cannot 'heave the lead' too often.

Nie. It is 'lead' indeed!

Min. Let us now return to our fixed Line: and this time we will take a point *not* on it, and through this point we will draw a second Line. You say that we can, if we choose, draw it in 'the same direction' as that of the first Line?

Nie. We do.

Min. In that case let me remind you of the warning I gave you a few minutes ago, that we have no geometrical meaning for the phrase 'the same direction,' *unless when used of Lines having a common point.* What geometrical meaning do you attach to the phrase when used of other Lines?

Nie. (*after a pause*) I fear we cannot give you a geometrical definition of it at present.

Min. No? Can you construct such Lines?

Nie. No, but really that is not necessary. We allow of 'hypothetical constructions' now-a-days.

Min. Well then, can you test whether a given Pair of Lines *have* this property ?   I mean, if I give you a certain diagram, and tell you its geometrical history, can you pronounce, on a certain Pair of finite Lines, which have no visible common point, as to whether they have this property ?

Nie. We cannot undertake it.

Min. You ask me, then, to believe in the reality of a class of ' Pairs of Lines ' possessing a property which you can neither define, nor construct, nor test ?

Nie. We can do none of these things, we admit: but yet the class is not quite so indefinite as you think.   We can give you a geometrical *description* of it.

Min. I shall be delighted to hear it.

Nie. We have agreed that a Pair of coincidental finite Lines have a certain relationship of direction, which we call ' the same direction,' and which you allow to be an intelligible geometrical relation ?

Min. Certainly.

Nie. Well, all we assert of this new class is that their relationship of direction is identical with that which belongs to coincidental Lines.

Min. It cannot be identical in *all* respects, for it certainly differs in this, that we cannot reach the conception of it by the same route.   I can form a conception of ' the same direction,' when the phrase is used of two Lines which have a common point, but it is only by considering that one ' falls on ' the other—that they have all other points common—that they coincide. When you ask me to form a conception of this relationship of direction, when asserted of other Lines, you know that none of these considerations will help me, and you do not furnish me with any substitutes for them. To me the relationship does *not* seem to be identical : I should prefer saying that separational lines have 'collateral,' or ' corresponding,' or ' separational' directions, to using the phrase 'the same direction' over again. It is, of course, true that 'collateral' directions produce the same results, as to angles made with a transversal, as 'coincidental' directions; but this seems to me to be a theorem, not an axiom.

Nic. You say that the relationship does *not* seem to you to be identical. I should like to know *where* you think you perceive any difference?

Min. I will try to make my meaning clearer by an illustration.

Suppose that I and several companions are walking along a railway, which will take us to a place we wish to visit. Some amuse themselves by walking on one of the rails ; some on another ; others wander along the line, crossing and recrossing. Now as we are all bound for the same place, we may say, roughly speaking, that we are *all* moving ' in the same direction': but that is speaking very roughly indeed. We make our language more exact, if we exclude the wanderers, and say that those who are walking along the rails are so moving. But it seems to me that our phrase becomes still more exact, if we limit it to those who are walking on one and the same rail.

As a second illustration, suppose two forces, acting on a

certain body ; and let them be equal in amount and opposite in direction.   Now, if they are acting along the same
line, we know that they neutralise each other, and that the
body remains at rest.   But if one be shifted ever so little
to one side, so that they act along parallel lines, then,
though still equal in amount and (according to the 'direction' theory) opposite in direction, they no longer neutralise
each other, but form a 'couple.'

As a third illustration, take two points on a certain
plane.   We may, first, draw a line through them and cause
them to move along that line : they are then undoubtedly
moving 'in the same direction.'   We may, secondly, draw
two lines through them, which meet or at least would meet
if produced, and cause them to move along those lines :
they are then undoubtedly moving ' in different directions.'
We may, thirdly, draw two parallel lines through them,
and cause them to move along those lines.   Surely this is
a new relationship of motion, not absolutely identical with
either of the former two ?   But if this new relationship be
not absolutely identical with that named 'in the same
direction,' it must belong to the class named 'in different
directions.'

Still, though this new relationship of direction is not
identical with the former in *all* respects, it is in *some* :
only, to prove this, we must use *some* disputed axiom, as
it will take us into Table II.   For instance, they are
identical as to angles made with transversals : this fact
is embodied in Tab. II. 3, viz. 'sT_m is real.'   Would you
like to adopt that, instead of 'sD is real'?

Nie. No. We are trying to dispense with Table II altogether.

Min. It is a vain attempt.

There is another remark I wish to make, before considering your second assertion. In asserting that there is a real class of non-coincidental Lines that have 'the same direction,' are you not also asserting that there is a real class of Lines that have no common point? For, if they *had* a common point, they must have ' *different* directions.'

Nie. I suppose we are.

Min. We will then, if you please, credit you with an axiom you have not expressed, viz. ' Lines may have no common point.' And here I must express an opinion that this ought to be *proved*, not assumed. Euclid has proved it in I. 27, which rests on no disputed axiom ; and I think it may be recorded as a distinct defect in your treatise, that you have assumed, as axiomatic, a truth which Euclid has *proved*.

My conclusion, as to this first assertion of yours, is that it is most decidedly *not* axiomatic.

Let us now consider your second assertion, that some non-coincidental Lines have ' different directions.' Here I must ask, as before, are you speaking of Lines which have a common point? If so, I am quite ready to grant the assertion.

Nie. Not exactly. It is rather a difficult matter to explain. The Lines we refer to *would*, as a matter of fact, meet if produced, and yet we do not suppose that fact known in speaking of them. What we ask you to believe is that there is a real class of non-coincidental finite Lines, which we do not yet know to have a common point, but which have 'different directions.' We shall assert presently, in another axiom, that such Lines will meet if produced; but we ask you to believe their reality independently of that fact.

Min. But the only geometrical meaning I know of, as yet, for the phrase 'different directions,' refers to Lines known to have a common point. What geometrical meaning do you attach to the phrase when used of other Lines?

Nie. We cannot define it.

Min. Nor construct it? Nor test it?

Nie. No.

Min. You ask me, then, to believe in the reality of *two* classes of 'pairs of Lines,' each possessing a property that you can neither define, nor construct, nor test?

Nie. That is true. But surely you admit the reality of the second class? Why, intersectional Lines are a case in point.

Min. Certainly. And so much I am willing to grant you. I allow that *some* non-coincidental Lines, viz. intersectional Lines, have 'different directions.' But as to 'the *same* direction,' you have given me no reason whatever for believing that there are *any* non-coincidental Lines which possess that property.

Nie. But surely there are two real distinct classes of non-coincidental Lines, 'intersectional' and 'separational'?

Min. Yes. Thanks to my interpolated axiom, you may now assume the reality of both.

Nie. And you will hardly assert that the relationship of direction, which belongs to a Pair of intersectional Lines, is

identical with that which belongs to a Pair of separational
Lines?

Min. I do not assert it.

Nie. And you allow that intersectional Lines have 'different directions'?

Min. Yes. Are you going to argue, from that, that
separational Lines must have 'the *same* direction'? Why
may I not say that intersectional Lines have *one* kind
of 'different directions' and that separational Lines have
another kind?

Nie. But *do* you say it?

Min. Certainly not. There is no evidence, at present,
one way or the other. For anything we know, separational Lines may all have 'the same direction,' or they
may all have 'different directions,' or there may be some
of each kind. I fear I must decline to grant the first
part of your axiom altogether, and the second part in
the sense of referring to Lines not known to have a
common point. You may now proceed.

NIEMAND *reads.*

P. 11. Ax. 7. 'Two different straight Lines which meet
one another have different directions.'

Min. That I grant you, heartily. It is, in fact, a
definition for 'different directions,' when used of Lines
which have a common point.

NIEMAND *reads.*

P. 11. Ax. 8. 'Two straight Lines which have different
directions would meet if prolonged indefinitely.'

Min. Am I to understand that, if we have before us a Pair of finite Lines which are not known to have a common point, but of which we *do* know that they have different directions, you ask me to believe that they will meet if produced ?

Nie. That is our meaning.

Min. We had better heave the lead once more, and return to our fixed Line, and a point not on it, through which we wish to draw a second Line. You ask me to grant that, if it be drawn so as to have a direction 'different' from that of the first Line, it will meet it if prolonged indefinitely ?

Nie. That is our humble petition.

Min. Will you be satisfied if I grant you that *some* Lines, so drawn, will meet the first Line ? *That* I would grant you with pleasure. I could draw millions of Lines which would fulfil the conditions, by simply taking points at random on the given Line, and joining them to the given point. Every Line, so constructed, would have a direction 'different' from that of the given Line, and would also meet it.

Nie. We will *not* be satisfied, even with millions! We ask you to grant that *every* Line, drawn through the given point with a direction 'different' from that of the given Line, will meet the given Line : and we ask you to grant this independently of, and antecedently to, any other information about the Lines except the fact that they have 'different' directions.

Min. But what meaning am I to attach to the phrase

'different directions,' independently of, and antecedently
to, the fact that they have a common point?

Nie. (*after a long silence*) I fear we can suggest none.

Min. Then I must decline to accept the axiom.

Nie. And yet this axiom is the converse of the preceding, which you granted so readily.

Min. The *technical* converse, my good sir, not the
logical! I will not suspect you of so gross a logical blunder
as the attempt to convert a universal affirmative *simpliciter*
instead of *per accidens*. The only converse, as you are no
doubt aware, to which you have any *logical* right, is
'*Some* Lines which have "different directions" would
meet if produced'; and *that* I grant you. It is true of
intersectional Lines, and I would limit the proposition *to*
such Lines, so that it would be equivalent to 'Lines
which would meet if produced would meet if produced'—
an indisputable truth, but *not* remarkable for novelty!
You may proceed.

Nie. I beg to hand in this diagram, and will read you
our explanation of it :—

'Thus *A* and *B* in the figure have the same direction;
and *C* and *D*, which meet, have different directions; and *E*
and *F*, which have different directions, would meet if produced far enough.'

Min. I grant the assertion about *C* and *D*; but I am
wholly unable to guess on what grounds you expect me

to grant that A and B 'have the same direction,' and
that E and F 'have different directions.' Do you expect
me to judge by eye? How if the lines were several yards
apart? Is *this* what Geometry is coming to? Proceed.

· NIEMAND *reads.*

Def. 19. 'Straight Lines which are not parts of the same
straight Line, but have the same direction, are called
parallels.'

Min. A *definition* is of course unobjectionable, since it
does not assert the *existence* of the thing defined : in fact,
it *asserts* nothing except the meaning which you intend to
attach to the word 'parallel.' But, as this word is used
in different senses, I will thank you to substitute for it,
in what you have yet to say about this matter, the phrase
' '*sD*'-related.'

But, before you read any more, let us get a clear idea
of your definition. We know of two real classes of Pairs
of Lines, 'coincidental' and 'intersectional'; and to these
we may (if we credit you with that unexpressed axiom,
'Lines may have no common point') add a third class,
which we may call 'separational.'

We also know that if a Pair of Lines has a common
point, and no separate point, it belongs to the first class;
if a common point, and a separate point, to the second.
Hence all Pairs of Lines, having a common point, must
belong to one or other of these classes. And since a Pair,
which has *no* common point, belongs to the third class,
we see that *every* conceivable Pair of Lines must belong
to one of these three classes.

This classification (by dichotomy) may be tabulated thus :—A Pair of Lines

which is $\left\{\begin{array}{l} \text{'}c\text{' and} \left\{\begin{array}{l} \text{'}s',\text{' is Coincidental.} \\ \text{'}s,\text{' .. Intersectional.} \end{array}\right. \\ \text{'}c',\text{' .... Separational.} \end{array}\right.$

We also know that——

Nie. (*sighing deeply*) You are heaving the lead again!

Min. I am : but we shall be in calmer water soon.

We also know that the ' Coincidental' class possesses two properties—'C' and 'D'; and that the ' Intersectional' class also possesses two properties—'X' and 'D'.'

Now if you choose to frame a definition by denying one property of each of these two classes, any Pair of Lines, so defined, is excluded from both of these classes, and must, *if it exists at all*, belong to the ' Separational' class. Remember, however, that you *may* have so framed your definition as to exclude your Pair of Lines from *existence*. For instance, if you choose to deny *both* the conditions of direction, and to say that Lines, which are 'DD',' are to be called so-and-so, you are simply describing a nonentity.

Nie. That is all quite clear.

Min. Your definition, then, amounts to this :—Lines, which are *not* coincidental, but which have the same direction, are said to be 'sD'-related.

Nie. It does.

Min. Well, here is another definition, which will answer your purpose just as well :—Lines, which are *not* intersectional, but which have different directions, are said to be '$X D'$'-related.

Nie. But I think I can prove to you that you have now done the very thing you cautioned me against: you have annihilated your Pair of Lines.

Min. That is a matter which we need not consider at present. Proceed.

NIEMAND *reads.*

P. 11. 'From this definition, and the axioms above given, the following results are immediately deduced :

(1) That parallel—I beg your pardon—that '*sD*'-related Lines would not meet however far they were produced. For if they met——'

Min. You need not trouble yourself to prove it. I grant that, *if* such Lines existed, they would not meet. Your assertion is simply the contranominal of Ax. 7, and therefore is necessarily true if the subject is real.

But remember that, though I have credited you with the axiom that, if we are given a Line and a point not on it, we can draw, through the point, *a* Line separational from the given Line, we do not yet know that it is *the only* such Line. *That* would take us into Table II. With our present knowledge, we must allow for the possibility of drawing any number of Lines through the given point, all separational from the given Line : and all I grant you is, that your ideal '*sD*'-related Line will, *if it exist at all,* be one of this group.

NIEMAND *reads.*

(2) 'That Lines which are '*sD*'-related to the same Line are '*sD*'-related to each other. For——'

Min. Wait a moment. I observe that you say that such Lines are 'sD'-related to each other. Might they not be 'cD'-related?

Nie. Certainly they might: but we do not wish to include that case in our predicate.

Min. Then you must limit your subject, and say '*Different* Lines.'

Nie. Very well.

Reads.

'That different Lines, which are 'sD'-related to the same Line, are 'sD'-related to each other. For they each have the same direction as that Line, and therefore the same direction as the other.'

Min. I am willing to grant you, without any proof, that, *if* such Lines existed, they would have the same direction with regard to each other. But you seem to have added a proof. Surely the second clause of the sentence is merely the first over again?

Nie. It *does* look like it, I admit.

Min. Then why insert 'for' and 'therefore,' as if it were a logical proof for the preceding clause?

Nie. Now I come to look at it, it is *not* a proof.

Min. But it professes to be one: I fear I must record it as an instance of '*Petitio principii.*' And then '*they each have*'! Is that good English?

Nie. We cannot defend it.

Min. As queer a specimen of logic, and of English composition, as I have seen for some time. You may proceed.

P. 12. Ax. 9. ' An angle may be conceived as transferred from one position to another, the direction of its arms remaining the same.'

Min. Let us first consider the right arm by itself. You assert that it may be transferred to a new position, its direction remaining the same ?

Nie. We do.

Min. You might, in fact, have here inserted an axiom ' A Line may be conceived as transferred from one position to another, its direction remaining the same ' ?

Nie. That would express our meaning.

Min. And this is virtually identical with your axiom ' Two different Lines may have the same direction ' ?

Nie. Certainly. They embody the same truth. But the one contemplates a single Line in two positions, and the other contemplates two Lines: the difference is very slight.

Min. Exactly so. Now let me ask you, do you mean, by the word ' angle,' a constant or a variable angle ?

Nie. I do not quite understand your question.

Min. I will put it more fully. Do you mean that the arms of the angle are rigidly connected, so that it cannot change its magnitude, or that they are merely hinged loosely together, as it were, so that it depends entirely on the relative motions of the two arms whether the angle changes its magnitude or not?

Nie. Why are we bound to settle the question at all ?

Min. I will tell you why. Suppose we say that the

arms are merely hinged together: in that case all you assert is that each arm may be transferred, its direction remaining the same; that is, you merely assert your 6th axiom twice over, once for the right arm and once for the left arm; and you do *not* assert that the angle will retain its magnitude. But in the theorem which follows, you clearly regard it as a constant angle, for you say 'the angle *AOD* would coincide with the angle *EKH. Therefore* the angle *AOD=EKH.*' But the 'therefore' has no force if *AOD* could change its magnitude. Thus you would be deducing, from an axiom where 'angle' is used in a peculiar sense, a conclusion in which it bears its ordinary sense. You have heard of the fallacy '*A dicto secundum Quid ad dictum Simpliciter*'?

Nie. (hastily) We are not going to commit ourselves to *that.* You may assume that we mean, by 'angle,' a rigid angle, which cannot change its magnitude.

Min. In that case you assert that, when a pair of Lines, terminated at a point, is transferred so that its vertex has a new position, these three conditions can be simultaneously fulfilled:—

(1) the right arm has 'the same direction' as before;

(2) the left arm has 'the same direction' as before;

(3) the magnitude of the angle is unchanged.

Nie. We do not dispute it.

Min. But any *two* of these conditions are sufficient, without the third, to determine the new state of things. For instance, taking (1) and (3), if we fix the position of the right arm, by giving it 'the same direction' as before, and also keep the magnitude of the angle unchanged, is not that enough to fix the position of the left arm, without mentioning (2)?

Nie. It certainly is.

Min. Your axiom asserts, then, that any two of these conditions lead to the third as a necessary result?

Nie. It does.

Min. Your axiom then contains *two* distinct assertions : the data of the first being (1) and (3) [or (2) and (3), which lead to a similar result], the data of the second being (1) and (2). These I will state as two separate axioms :—

9 (*a*). If a Pair of Lines, terminated at a point, be transferred to a new position, so that the direction of one of the Lines, and the magnitude of the included angle, remain the same; the direction of the other Line will remain the same.

9 (*β*). If a Pair of Lines, terminated at a point, be transferred to a new position, so that their directions remain the same; the magnitude of the included angle will remain the same.

Have I represented your meaning correctly?

Nie. We have no objection to make.

Min. We will return to this subject directly. I must now ask you to read the enunciation of Th. 4, omitting, for simplicity's sake, all about supplementary angles, and assuming the Lines to be taken 'the same way.'

NIEMAND *reads.*

P. 12, Th. 4. 'If two Lines are respectively '*sD*'-related to two other Lines, the angle made by the first Pair will be equal to the angle made by the second Pair.'

Min. The '*s*' is of course superfluous, for if the Lines
are '*cD*', it is equally true.   May I re-word it thus?—

'If two Pairs of Lines, each terminated at a point, be
such that the directions of one Pair are respectively the
same as those of the other; the included angles are
equal'.

Nie. Yes, if you like.

Min. But surely the only difference, between Ax. 9 (β)
and this, is that in the axiom we contemplated a single
Pair of Lines transferred, while *here* we contemplate two
Pairs?

Nie. That is the only difference, we admit.

Min. Then I must say that it is anything but good
logic to take two propositions, distinguished only by a
trivial difference in form, and to call one an axiom and
the other a theorem deduced from it!   A very gross case
of '*Petitio principii*,' I fear!

Nie. (*after a long pause*) Well! We admit that it is
not exactly a theorem : it is only a new form of the axiom.

Min. Quite so : and as it is a more convenient form for
my purpose, I will with your permission adopt it as a substitute for the axiom.   Now as to the corollary of this
theorem: *that*, I think, is merely a particular case of
Ax. 9 (β), one of the arms being slid along the infinite
Line of which it forms a part, and thus of course having
'the same direction' as before?

Nie. It is so.

Min. And, as this is a more convenient form still,
I will restate your assertions, limiting them to this particular case :—

Ax. 9 (*a*). Lines, which make equal corresponding angles with a certain transversal, have the same direction.

Ax. 9 (*β*). Lines, which have the same direction, make equal corresponding angles with any transversal.

Am I right in saying that these two assertions are virtually involved in your axiom?

Nie. We cannot deny it.

Min. Now in 9 (*a*) you ask me to believe that Lines possessing a certain geometrical property, which can be defined, constructed, and tested, possess also a property which, in the case of different Lines, we can neither define, nor construct, nor test. There is nothing axiomatic in this. It is much more like a definition of '*D*' when asserted of different Lines, for which we have as yet no definition at all. Will you not permit me to insert it, as a definition, before Ax. 6? We might word it thus :—

'If two different Lines make equal angles with a certain transversal, they are said to have the same direction: if unequal, different directions.'

This interpolation would have the advantage of making Ax. 6 (which I have hitherto declined to grant) indisputably true.

Nie. (after a pause) No. We cannot adopt it as a definition so early in the subject.

Min. You are right. You probably saw the pitfall which I had ready for you, that this same definition would make your 8th axiom exactly equivalent to Euclid's 12th! From this catastrophe you have hitherto been saved solely by the absence of geometrical meaning in your

phrase 'the same direction,' when applied to different
Lines.   Once define it, and you are lost!

Nie. We are aware of that, and prefer all the inconvenience which results from the absence of a definition.

Min. The 'inconvenience,' so far, has consisted of the
ruin of Ax. 6 and Ax. 8.   Let us now return to Ax. 9.

As to 9 (β), it is of course obviously true with regard
to *coincidental* Lines : with regard to *different* 'Lines,
which have the same direction,' I grant you that, *if* such
Lines existed, they *would* make equal corresponding angles
with any transversal ; for they would then have a relationship of direction identical with that which belongs to coincidental Lines.   But all this rests on an 'if'—*if* they
existed.

Now let us combine 9 (β) with Axiom 6, and see what
it is you ask me to grant.   It is as follows :—

'There can be a Pair of different Lines that make equal
angles with *any* transversal.'

I am not misrepresenting you, I think, if I say that
you propound this as axiomatic truth—which, I need
hardly remark, lands us in Table II, being the proposition 'sT_m is real.'

Nie. We accept the responsibility of the two axioms
separately, but not of a logical deduction from the two.

Min. There are certainly *some* logical deductions from
axioms (contranominals for instance) that are not so
axiomatic as the axioms from which they come : but
surely if you tell me 'it is axiomatic that X is Y' and
'it is axiomatic that Y is Z,' it is much the same as saying
'it is axiomatic that X is Z'?

Nie. It is very like it, we admit.

Min. Now take one more combination. Take 9 (a) and 9 (β). We thus eliminate the mysterious property altogether, and get a proposition whose subject and predicate are perfectly definite geometrical conceptions—a proposition which you assert to be, if not perfectly axiomatic, yet so nearly so as to be easily deducible from two axioms—a proposition which again lands us in Table II, and which, I will venture to say, is less axiomatic than any axiom that has ever yet been suggested for the acceptance of the human race. We get *this*:—

'Lines which make equal corresponding angles with a certain transversal do so with *any* transversal,' which is Tab. II. 4. ' st_m ℙ T_m.'

Here we have, condensed into one appalling sentence, the whole substance of Euclid I. 27, 28, and 29 (for the fact that the lines are 'separational' is merely used by him as a go-between). Here we have the whole difficulty of Parallels swallowed at one gulp. Why, Euclid's much-abused 12th axiom is nothing to it! If we had (what I fear has yet to be discovered) a unit of 'axiomaticity,' I should expect to find that Euclid's 12th axiom (which you call in your Preface, at p. xiii, 'not axiomatic') was twenty or thirty times as axiomatic as this! I need not ask you for any further proof of Euc. I. 32. This wondrous axiom, or quasi-axiom, is quite sufficient machinery for your purpose, along with Euc. I. 13, which of course we grant you. Have you thought it necessary to provide any other machinery?

Nie. No.

Min. Euclid requires, besides I. 13, the following machinery :—Props. 4, 15, 16, 27, 28, Ax. 12, and Prop. 29. And for all this you offer, as a sufficient substitute, one single axiom!

Nic. Two, if you please. You are forgetting Ax. 6.

Min. No, I repeat it—one single axiom. Ax. 6 is contained in Ax. 9 (*a*): when the subject is known to be real, the proposition necessarily asserts the reality of the predicate.

Nie. That we must admit to be true.

Min. I need hardly say that I must decline to grant this so-called ‘Axiom,’ even though its collapse should involve that of your entire system of ‘Parallels.’ And now that we have fully discussed the subject of direction, I wish to ask you one question which will, I think, sum up the whole difficulty in a few words. It is, in fact, *the* crucial test as to whether ‘direction’ is, or is not, a logical method of proving the properties of parallels.

You assert, as axiomatic, that different Lines exist, whose relationship of direction is identical with that which exists between coincidental Lines.

Nie. Yes.

Min. Now, does the phrase ‘the same direction,’ when used of two Lines not known to have a common point, convey to your mind a clear geometrical conception?

Nie. Yes, we can form a clear idea of it, though we cannot define it.

Min. And is that idea (this is the crucial question) *independent of all subsequent knowledge of the properties of Parallels?*

Nie. We believe so.

Min. Let us make sure that there is no self-deception in this. You feel certain you are not unconsciously picturing the Lines to yourself as being equidistant, for instance?

Nie. No, they suggest no such idea to us. We introduce the idea of equidistance later on in the book, but we do not feel that our first conception of 'the same direction' includes it at all.

Min. I think you are right, though Mr. Cuthbertson, in his 'Euclidian Geometry,' says (Pref. p. vi.) 'the conception of a parallelogram is not that of a figure whose opposite sides will never meet . . . . , but rather that of a figure whose opposite sides are equidistant.' But do you feel equally certain that you are not unconsciously using your subsequent knowledge that Lines exist which make equal angles with all transversals?

Nie. We are not so clear about *that*. It is, of course, extremely difficult to divest one's mind of all later knowledge, and to place oneself in the mental attitude of one who is totally ignorant of the subject.

Min. Very difficult, no doubt, but absolutely essential, if you mean to write a book adapted to the use of beginners. My own belief as to the course of thought needed to grasp the theory of 'direction' is this:—first you grasp the idea of 'the same direction' as regards Lines which have a common point; next, you convince yourself, by some *other* means, that different Lines exist which make equal angles with all transversals; thirdly, you go back, armed with this new piece of knowledge, and use it unconsciously, in forming an idea of 'the same direction' as regards

different Lines.   And I believe that the course of thought
in the mind of a beginner is simply this :—he grasps,
easily enough, the idea of 'the same direction' as regards
Lines which have a common point; but when you put
before him the idea of *different* Lines, and ask him to
realise the meaning of the phrase, when applied to such
Lines, he, finding that the former geometrical conception
of 'coincidence' is not applicable in this case, and knowing nothing of the idea, which is latent in *your* mind,
of Lines which make equal angles with all transversals,
simply fails to attach *any* idea at all to the phrase, and
accepts it blindly, from faith in his teacher, and is from
that moment, until he reaches the theorem about transversals, walking in the dark.

Nie. If this is true, of course the theory of 'direction,'
however beautiful in itself, is not adapted for purposes of
teaching.

Min. That is my own firm conviction.   But I fear I
may have wearied you by discussing this matter at such
great length.   Let us turn to another subject.   What is
your practical test for knowing whether two finite Lines
will meet if produced ?

Nie. You have already heard our 8th Axiom (p. 11).
'Two straight lines which have different directions would
meet if produced.'

Min. But, even if that were axiomatic (which I deny), it
would be no *practical* test, for you have admitted that you
have no means of knowing whether two Lines, not known
to have a common point, have or have not different directions.

Nie. We must refer you to p. 14. Th. 5. Cor. 2, where we prove that Lines, which make equal angles with a certain transversal, have the same direction.

Min. Which you had already asserted, if you remember, in Ax. 9.

Nie. Well then, we refer you to Ax. 9 as containing the same truth.

Min. And having got that truth, whether lawfully or not, what do you do with it?

Nie. Why, surely it is almost the same as saying that, if they make *un*equal angles, they have *different* directions.

Min. And what then?

Nie. Then, combining this with the axiom you refused to grant, namely, that Lines having different directions will meet, we get a practical test, such as you were asking for.

Min. (dreamily) I see! You get rid of the 'different directions' altogether, and the result is that ' Lines, which make unequal angles with a certain transversal, will meet if produced,' which is Tab. II. 2 (2). '$st_{m'}$ ⚑ X.' And this you assert as axiomatic truth?

Nie. (uneasily) Yes.

Min. Surely I have read something like it before? Could it have been Euclid's 12th Axiom? And have I not somewhere read words like these :—' Euclid's treatment of parallels distinctly breaks down in Logic. It rests on an axiom which is not axiomatic '?

Nie. We have nowhere *stated* this axiom which you put into our mouth.

Min. No? Then how, may I ask, do you prove that

particular Lines *will* meet? You *must* have to prove it sometimes, you know.

Nie. We have not had to prove it anywhere, that we are aware of.

Min. Then there must be some gaps in your arguments. Let us see. Please to turn to p. 46. Prob. 7. Here you make, at the ends of a Line *CD*, angles equal to two given angles (which, as you tell us below, ' must be together less than two right angles '), and you then say ' let their sides meet in *O.*' How do you know that they *will* meet?

Nie. You have found *one* hiatus, we grant. Can you point out another in the whole book?

Min. I can. At p. 70 I find the words ' Join *QG*, and produce it to meet *FH* produced in *S.*' And again at p. 88. ' Hence the centre must be at *O*, the point of intersection of these perpendiculars.' In both these cases I would ask, as before, how do you know that the Lines in question *will* meet?

Nie. We had not observed the omissions before, and we must admit that they constitute a serious hiatus.

Min. A most serious one. A student, who had been taught such proofs as these, would be almost sure to try the plan in cases where the Lines would *not* really meet, and his assumption would lead him to results more remarkable for novelty than truth.

Let us now take a general survey of your book. And First, as to the propositions of Euclid which you omit—

Nie. You are alluding to Prop. 7, I suppose. Surely its only use is to prove Prop. 8, which we have done very well without it.

Min. That is quite a venial omission. The others that I miss are 27, 28, 29, 30, 33, 34, 35, 36, 37, 38, 39, 40, 41, and 43 : rather a formidable list.

Nie. You are much mistaken! Nearly all of those are in our book, or could be deduced in a moment from theorems in it.

Min. Let us take I. 34 as an instance.

Nie. *That* we give you, almost in the words of Euclid, at p. 37.

Reads.

Th. 22. 'The opposite angles and sides of a parallelogram will be equal, and the diagonal, or the Line which joins its opposite angles, will bisect it.'

Min. Well, but *your* parallelogram is not what Euclid contemplates. *He* means by the word that the opposite sides are separational—a property whose reality he has demonstrated in I. 27; whereas *you* mean that they have the same direction—a property whose reality, when asserted of different Lines, has nowhere been satisfactorily proved.

Nie. We have proved it at p. 14. Th. 5. Cor. 2.

Min. Which, if traced back, will be seen to depend ultimately on your 6th Axiom, where you *assume* the reality of such Lines. But, even if your Theorem *had* been shown to refer to a real figure, how would that prove Euc. I. 34 ?

Nie. You only need the link that separational Lines have the same direction.

Min. Have you supplied that link ?

Nie. No : but the reader can easily make it for himself. It is the 'contranominal' (as you call it) of our 8th Axiom, 'two straight Lines which have different directions would meet if prolonged indefinitely.'

Min. Your pupils must be remarkably clever at drawing deductions and filling up gaps in an argument, if they usually supply that link, as well as the proof that separational Lines exist at all, for themselves. But, as *you* do not supply these things, it seems fair to say that your book omits all the propositions which I have enumerated.

I will now take a general survey of your book, and select a few points which seem to call for remark.

At p. 9, we have a deduction, from a Definition and an Axiom, which involves the Fallacy '*Illicit process of the Minor.*' The passage is as follows :—

Def. 11. 'A straight line is said to be *perpendicular* to another straight Line when it makes a right angle with it. Hence there can be only one perpendicular to a given Line at a given point, on one side of that Line, because only one Line can make a right angle with the given Line at that point.'

Thrown into syllogistic form, the argument may be stated thus :—

'All Lines drawn at right angles to a given Line at a given point, on one side of it, are coincident;

all Lines drawn at right angles to a given Line &c. are *perpendiculars* to that Line &c. ;

therefore, all perpendiculars to a given Line &c. are coincident.' That is, 'All X is Y; all X is Z; therefore all Z is Y.'

Or we may simply exhibit the fallacy by parodying the argument in other subject-matter.

'A Line is said to be *acutely inclined* to the sides of a triangle when it bisects the vertical angle. Hence there can be only one Line acutely inclined to the sides of a triangle, because only one Line can bisect the vertical angle.' You cannot deny, I think, that this argument is, in *form*, as valid as yours?

Nie. We cannot deny it.

MINOS *reads.*

P. 14. Th. 5, Cor. 1. ' Hence if two straight Lines which are not parallel are intersected by a third, the alternate angles will be not equal, and the interior angles on the same side of the intersecting Line will be not supplementary.' Excuse the apparent incivility of the remark, but this Corollary is false.

Nie. You amaze me !

Min. You have simply to take, as an instance, a ·Pair of *coincidental* Lines, which most certainly answer to your description of ' not parallel.'

Nie. It is an oversight.

Min. So I suppose : it is a species of literary phenomenon in which your Manual is rich.

Your proof of Cor. 2. is a delicious collection of negatives.

Reads.

' Cor. 2. *Hence also if the corresponding angles are equal, or the alternate angles equal, or the interior angles supplementary, the Lines will be parallel.*

'For they can*not* be *not* parallel, for then the cor-responding and alternate angles would be *un*equal by Cor. 1.'

Should I be justified in calling this a somewhat *knotty* passage?

Nie. You have no right to make such a remark. It is a mere jest!

Min. Well, we will be serious again.

At p. 9, you stated *more* than the data authorised: we now come to a set-off against this, since we shall find you asserting *less* than you ought to do. I will read the passage :—

P. 26, Th. 15. 'If two triangles are equiangular to one another and have a side of the one equal to the correspond-ing side of the other, the triangles will be equal in all respects.'

This contains a superfluous *datum*: it would have been enough to say 'if two triangles have two angles of the one equal to two angles of the other &c.'

Nie. Well, it is at worst a superfluity : the enunciation is really identical with Euclid's.

Min. By no means. The logical effect of a superfluous *datum* is to *limit* the extent of a proposition : and, if the proposition be 'universal,' it reduces it to 'particular'; i.e. it changes 'all A is B' into 'some A is B.' For suppose we take the proposition 'all A is B,' and sub-stitute for it 'all that is both A and X is B,' we may be *accidentally* making an assertion of the same extent as before, for it may happen that the whole class 'A' pos-sesses the property 'X'; but, so far as logical *form* is concerned, we have reduced the proposition to ' *some* things that are A (viz. those which are also X) are B.'

I turn now to p. 27, where I observe a new proof for Euc. I. 24.

Nie. New and, we hope, neat and short.

Min. Charmingly neat and short, *as it stands* : but this method really requires the discussion of *five* cases, each with its own figure.

Nie. How do you make that out ?

Min. The five cases are :—

(1) Vertical angles together less than two right angles, and adjacent base angles acute (the case you give).

(2) Adjacent base-angles right.

(3) Adjacent base-angles obtuse.

These two cases are proved along with the first.

(4) Vertical angles together equal to two right angles.

This requires a new proof, as we must substitute for the words ' the bisector of the angle FAC,' the words ' the perpendicular to FC drawn through A.'

(5) Vertical angles together greater than two right angles.

This also requires a new proof, as we must insert, after the words ' the bisector of the angle FAC,' the words ' produced through A,' and must then prove (by your Th. 1) that the angles OAC, OAF, are equal.

On the whole, I take this to be the most cumbrous proof yet suggested for this theorem.

We now come to what is probably the most extraordinary Corollary ever yet propounded in a geometrical treatise. Turn to pages 30 and 31.

Th. 20. 'If two triangles have two sides of the one
equal to two sides of the other, and the angle opposite
that which is not the less of the two sides of the one
equal to the corresponding angle of the other, the triangles
shall be equal in all respects.

'Cor. 1. If the side opposite the given angle were less
than the side adjacent, there would be two triangles, as in
the figure; and the proof given above is inapplicable.

'This is called *the ambiguous* case.'

The whole proposition is a grand specimen of obscure
writing and bad English, 'is' and 'are,' 'could,' and
'would,' alternating throughout with the most charming
impartiality: but what impresses me most is the probable
effect of this wondrous Corollary on the brain of a simple
reader, coming breathless and exhausted from a death-
struggle with the preceding theorem. I can imagine him
saying wildly to himself ' If two triangles fulfil such and
such conditions, such and such things follow: but, if one
of the conditions were to fail, *there would be two triangles!*
I must be dreaming! Let me dip my head in cold water,
and read it all again. If two triangles . . . there *would* be
two triangles. Oh, my poor brains!'

Nie. You are pleased to be satirical: it *is* rather obscure
writing, we confess.

Min. It is indeed! You do well in calling it *the ambiguous case*.

At p. 33, I see the heading 'Theorems of equality':
but you only give *two* of them, the second being 'the
bisectors of the three angles of a triangle meet in one
point,' which, as a specimen of 'theorems of equality,'

is probably unique in the literature of Geometry.   I cannot wonder at your not attempting to extend the collection.

At p. 40 I read, 'It is assumed here that if a circle has one point inside another circle, the circumferences will intersect one another.'   This I believe to be the boldest assumption yet made in Modern Geometry.

At pp. 40, 42 you assume a length 'greater than half' a given Line, without having shewn how to bisect Lines. Two cases of '*Petitio Principii.*'   (See p. 58.)

P. 69. Here we have a problem (which you call 'the quadrature of a rectilineal area') occupying three pages and a half.   It is 'approached' by four 'stages,' which is a euphemism for saying that this fearful proposition contains *four* of Euclid's problems, viz. I. 42, 44, 45, and II. 14.

P. 73. 2. 'Find a point equally distant from three given straight lines.'   Is it fair to give this without any limitation?   What if the given lines were parallel?

P. 84. 'If A, B, C . . . as conditions involve D as a result, and the failure of C involves a failure of D; then A, B, D . . . as conditions involve C as a result.'   If not-C proves not-D, then D proves C. A and B are irrelevant and obscure the statement.   I observe, in passing, the subtle distinction which you suggest between '*the* failure of C' and '*a* failure of D.' D is a habitual bankrupt, who has often passed through the court, and is well used to failures: but, when C fails, his collapse is final, and 'leaves not a wrack behind'!

P. 90. 'Given a curve, to ascertain whether it is an arc

of a circle or not.' What does 'given a curve' mean? If it means a line drawn with ink on paper, we may safely say at once 'it is *not* a circle.'

P. 96. Def. 15. 'When one of the points in which a secant cuts a circle is made to move up to, and ultimately coincide with, the other, the ultimate position of the secant is called *the tangent* at that point.' (The idea of *the position of a Line* being itself a Line is queer enough: I suppose you would say '*the ultimate position* of Whittington was the Lord Mayor of London.' But this is by the way: of course you mean 'the secant in its ultimate position.') Now let us take three points on a circle, the middle one fixed, the others movable; and through the middle one let us draw two secants, each passing through one of the other points; and then let us make the other points 'move up to, and ultimately coincide with,' the middle one. We have no ground for saying that these two secants, in their ultimate positions, will coincide. Hence the phrase '*the* tangent' assumes, without proof, Th. 7. Cor. 1, viz. 'there can be only one tangent to a circle at a given point.' This is a '*Petitio principii.*'

P. 97. Th. 6. The secant consists of two portions, each terminated at the fixed point. All that you prove here is that the portion which has hitherto cut the circle is ultimately outside: and you jump, without a shadow of proof, to the conclusion that the same thing is true of the *other* portion! Why should not the second portion begin to cut the circle at the precise moment when the first ceases to do so? This is another '*Petitio Principii.*'

P. 129, line 3 from end. 'Abstract quantities are the

means that we use to express the concrete.' Excluding
such physical 'means' as pen and ink or the human voice
(to which you do not seem to allude), I presume that the
'means' referred to in this mysterious sentence are 'pure
numbers.' At any rate the only instances given are 'seven,
five, three.' Now take P. 130, l. 5, '*Abstract quantities*
and *ratios* are precisely the same things.' Hence all ratios
are numbers. But in the middle of the same page we
read that 'all numbers are ratios, but all ratios are *not*
numbers.' I leave this without further remark.

I will now sum up the conclusions I have come to with
respect to your Manual.

(1) As to 'straight Lines' you suggest a useful extension of Euclid's axiom.

(2) As to angles and right angles, your extension of
the limit of size is, in my opinion, objectionable. In
other respects your language, though hazy, agrees on the
whole with Euclid.

(3) As to 'parallels,' there is a good deal to be said, and
that not very flattering, I fear.

In Ax. 6, you assert the reality of different Lines
having the same direction—a property you can neither
define, nor construct, nor test.

You also assert (by implication) the reality of separational Lines, which Euclid *proves.*

You also assert the reality of Lines, not known to have

a common point, but having different directions—a property you can neither define, nor construct, nor test.

In Ax. 8, you assert that the undefined Lines last mentioned would meet if produced.

These axioms, therefore, are not axiomatic.

In proving result (2), you are guilty of the fallacy ' *Petitio principii.*'

In Ax. 9 and Th. 4 taken together, if the word 'angle' in Ax. 9 means 'variable angle,' you are guilty of the fallacy ' *A dicto secundum Quid ad dictum simpliciter*'; if ' constant angle,' of the fallacy ' *Petitio Principii.*'

In Ax. 9 (*a*), you assert that Lines possessing a certain real geometrical property, viz. making equal angles with a certain transversal, possess also the before-mentioned undefined property. This is not axiomatic.

In Ax. 9 (*β*) combined with Ax. 6, you assert the reality of Lines which make equal angles with all transversals. This is not more axiomatic than Euc. Ax. 12.

In Ax. 9 (*a*) combined with Ax. 9 (*β*), you assert that Lines, which make equal angles with a certain transversal, do so with all transversals. This I believe to be the most unaxiomatic axiom ever yet proposed.

(4) You furnish no practical test for the meeting of finite Lines, and consequently you never prove (however necessary for the matter in hand) that any particular Lines *will* meet. And when we come to examine what practical test can possibly be extracted from your axioms, the only result is an imperfect edition of Euclid's 12th axiom !

The sum total of the chief defects which I have noticed is as follows :—

> fourteen of Euclid's Theorems in Book I. omitted ;
> seven unaxiomatic axioms ;
> six instances of ' *Petitio Principii* ' ;
> one instance of ' *Illicit Process of the Minor.*'

The abundant specimens of logical inaccuracy, and of loose writing generally, which I have here collected would, I feel sure, in a mere popular treatise be discreditable—in a scientific treatise, however modestly put forth, deplorable—but in a treatise avowedly put forth as a model of logical precision, and *intended to supersede Euclid*, they are simply monstrous.

My ultimate conclusion on your Manual is that it has *no claim whatever* to be adopted as *the* Manual for purposes of teaching and examination.

ACT II.

SCENE V.

§ 2. PIERCE.

'dum brevis esse laboro,
Obscurus fio.'

Nie. I lay before you '*An Elementary Treatise on Plane and Solid Geometry*' by BENJAMIN PIERCE, A.M., Perkins Professor of Astronomy and Mathematics in Harvard University, published in 1872.

Min. As I have already considered, at great length, the subject of direction as treated by Mr. Wilson, I need not trouble you as to any matters where Mr. Pierce's treatment does not materially differ from his. Is there any material difference in the treatment of a straight line?

Nie. He has a definition of direction which will, I think, be new to you:—

Reads.

P. 5, § 11, Def. 'The *Direction of a Line* in any part is the direction of a point at that part from the next preceding point of the line.'

Min. That sounds mysterious.   Which way along a line are ' preceding ' points to be found ?

Nie. Both ways.   He adds, directly afterwards, ' a Line has two different directions,' etc.

Min. So your Definition needs a postscript ?   That is rather clumsy writing.   But there is yet another difficulty.   How far from a point is the ' next ' point ?

Nie. At an infinitely small distance, of course.   You will find the matter fully discussed in any work on the Infinitesimal Calculus.

Min. A most satisfactory answer for a teacher to make to a pupil just beginning Geometry !   I see nothing else to remark on in your treatment of the Line, except that you state, as an axiom, that 'a straight Line is the shortest way from one point to another.'   I have already given, in my review of M. Legendre, my reasons for thinking that this is not a fair axiom, and ought to be a theorem (see p. 55).

There is nothing particular to notice in your treatment of angles and right angles.   Let us go on to Parallels.   How do you prove Euc. I. 32 ?

NIEMAND *reads.*

P. 9, § 27, Def. *' Parallel* Lines are straight Lines which have the same Direction.'

Min. I· presume you do not mean to include coincidental Lines ?

Nie. Certainly not.   We see the omission.   Allow us to insert the word ' different.'

Min. Very well.   Then your definition combines the two

properties 'different' and 'having the same direction.'
Bear in mind that you have yet to prove the *reality* of
such Lines. And may I request you in future to call
such Lines '*sD*'-related? But if you wish to assert any
thing of them which is also true of coincidental Lines,
you had better drop the '*s*,' and simply call them
'Lines which have the same direction,' so as to include
both classes.

Nie. Very well.

NIEMAND *reads.*

P. 9. § 28. Th. '*sD*'-related Lines cannot meet, however far they are produced.'

Min. Or rather '*could* not meet, if they existed.' Proceed.

NIEMAND *reads.*

P. 9. § 29. Th. 'Two angles are equal, when their sides
have the same direction.'

Min. How do you define 'same direction' for different
Lines?

Nie. We cannot define it.

Min. Then I cannot admit that such Lines exist. But
even if I *did* admit their reality, why should the angles
be equal?

Nie. Because 'the difference of direction' is the same
in each case.

Min. But how would that prove the angles equal?
Do you define 'angle' as the 'difference of direction' of
two lines?

Nie. Not exactly. We have stated (p. 6, § 19) 'The

magnitude of the angle depends solely upon the *difference of direction* of its sides at the vertex.'

Min. But the difference of direction also possesses 'magnitude.' Is that magnitude a wholly *in*dependent entity? Or does it, in its turn, depend to some extent upon the *angle*? Seriously, all these subtleties must be very trying to a beginner. But we had better proceed to the next theorem. I am anxious to see where, in this system, these creatures of the imagination, these '*sD*'-related Lines, are to appear as actually existent.

Nie. We next prove (p. 9. § 30) that Lines, which have the same direction, make equal angles with all transversals.

Min. That is merely a particular case of your last theorem.

Nie. And then that two Lines, which make equal angles with a transversal, have the same direction.

Min. Ah, *that* would bring them into existence at once! Let us hear the proof of that.

Nie. The proof is that if, through the point where the first Line is cut by the transversal, a Line be drawn having the same direction as the second, it makes equal angles with the transversal, and therefore coincides with the first Line.

Min. You assume, then, that a Line *can* be drawn through that point, having the same direction as the second Line?

Nie. Yes.

Min. That is, you assume, without proof, that different Lines can have the same direction. On the whole, then, though Mr. Pierce's system differs slightly from Mr.

Wilson's, both rest on the same vicious axiom, that different Lines *can* exist, which possess a property called 'the same direction'—a phrase which is intelligible enough when used of two Lines which have a common point, but which, when applied to two Lines *not* known to have a common point, can neither be defined, nor constructed. We need not pursue the subject further. Have you provided any test for knowing whether two given finite Lines will meet if produced?

Nie. We have not thought it necessary.

Min. Then the only other remark I have to make on this singularly compendious treatise is that, of the 35 theorems which Euclid gives us in his First Book, it reproduces just sixteen : the omissions being 16, 17, 25, 26 (2), 27 and 28, 29, 30, 33, 34, 35, 36, 37, 38, 39, 40, 41, 43, 47, and 48.

Nie. Most of those are in the book. For example, § 30 answers to Euc. I. 29.

Min. Only by proving that separational Lines have the same direction : which you have not done.

Nie. At any rate we have Euc. I. 47 in our § 256.

Min. Oh, no doubt! Long after going through ratios, which necessarily include incommensurables ; and long after the axiom (§ 99) 'Infinitely small quantities may be neglected'! No, no: so far as *beginners* are concerned, there is no Euc. I. 47 in *this* book!

My conclusion is that, however useful this Manual may be to an advanced student, it is *not* adapted to the wants of a beginner.

ACT II.

Scene V.

§ 3. Willock.

'This work . . . . no doubt, has its faults.'
Willock, *Pref.* p. 1.

Nie. I lay before you '*The Elementary Geometry of the Right Line and Circle*' by W. A. Willock, D.D., formerly Fellow of Trinity College, Dublin, published in 1875.

Min. I have gone through the subject of ' direction ' so minutely in reviewing Mr. Wilson's book, that I need not discuss with you any points in which your client essentially agrees with him. We may, I think, pass over the subject of the Right Line altogether?

Nie. Yes.

Min. And as to Angles and Right Angles, I see no novelty in Dr. Willock's book, except that he defines an Angle as ' the divergence ·of two directions,' which is virtually the same as Euclid's definition.

Nie. That I think is all.

Min. Then we can proceed at once to the subject of Parallels. Will you kindly give me your proof of Euc. I. 32 from the beginning?

Niemand reads.

P. 10, Th. 1. '*Two Directives can intersect in only one point.*'

Min. By 'Directive' you mean an 'infinite Line'?

Nie. Yes.

Min. Well, I need hardly trouble you to prove it as a theorem, being quite willing to grant it as an axiom. What is the next theorem?

Niemand reads.

P. 11, Th. 5. '*Parallel Directives cannot meet.*'

Min. We will call them '*sD*'-related, if you please. I grant it, provisionally. *If* such Lines exist, they cannot meet.

Niemand reads.

P. 11, Th. 7. '*Only one Line, 'sD'-related to a Directive, can be drawn through a point.*'

Min. Does that assert that one *can* be drawn? Or does it simply deny the possibility of drawing *two*?

Nie. The *proof* only applies to the denial: but the assertion is certainly involved in the enunciation. At all events, if not assumed here, it *is* assumed later on.

Min. Then I will at this point credit you with *one* unwarrantable axiom, namely, that different Lines can have the same direction. The theorem itself I grant.

NIEMAND *reads.*

P. 12, Th. 8. ' *The angles of intersection of a Transversal with two* ' *sD*'-*related Directives are equal*.'

Min. Do you prove that by Mr. Wilson's method?

Nie. Not quite. *He* does it by transferring an angle: *we* do it by divergence of directions.

Min. I prefer *your* method. All it needs to make it complete is the proof of the reality of such Lines: but *that* is unattainable, and its absence is fatal to the whole system. Nay, more: the fact, that the reality of such Lines leads by a logical necessity to the reality of Lines which make equal angles with *any* transversal, reacts upon that unfortunate axiom, and destroys the little hope it ever had of being granted without proof. In point of fact, in asking to have the axiom granted, you were virtually asking to have this other reality granted as axiomatic—but all this I have already explained (p. 100).

NIEMAND *reads.*

P. 13, Th. 10. ' *If a Transversal cut two Directives and make the angles of intersection with them equal, the Directives are* ' *sD*'-*related*.'

Min. The subject of your proposition is indisputably real. If then you can prove this theorem, you will thereby prove the reality of '*sD*'-related Lines. But I fear you have assumed it already in Th. 7. There is still, however, a gleam of hope: perhaps you do not need Th. 7 in proving this?

Nie. We do not : but I fear that will not mend matters, as we assume, in the course of this theorem, that a Line can be drawn through a given point, so as to have the same direction as a given Line.

Min. Then we need not examine it further : it must perish with the faulty axiom on which it rests. What is your next theorem ?

Nie. It answers to Euc. I. 16, 17, and is proved by the theorem you have just rejected.

Min. Then I must reject its *proof*, but I will grant you the theorem itself, if you like, as we know it *can* be proved from undisputed axioms. What comes next ?

Niemand reads.

P. 14, Th. 13. ' *If a Transversal meet two Directives, and make angles with them, the External greater than the Internal, or the sum of the two Internal angles less than two right angles, the two directives must meet.*'

Min. A proof for Euclid's Axiom ? That is interesting.

Niemand reads.

' For, suppose they do not meet. Then, they should be ' *sD* '-related——'

Min. (*interrupting*) '*Should* be '*sD*'-related' ? Does that mean that they *are* '*sD*'-related ?

Nie. Yes, I think so.

Min. That is, you assume that separational Lines have the same direction ?

Nie. We do.

Min. A fearful assumption ! (*A long silence*) Well ?

Nie. I am waiting to know whether you grant it.

Min. Unquestionably not! I must mark it against you as an axiom of the most monstrous character! Mr. Wilson himself does not assume this, though he *does* assume its contranominal, that Lines having different directions will meet (see p. 88). And what I said then I say now—unaxiomatic! But supposing it granted, how would you prove the theorem?

NIEMAND *reads.*

'Then, they should be '*sD*'-related; and the external angle should be equal (Th. 8) to the internal; which is contrary to the supposition.'

Min. Quite so. But Th. 8, which you quote, itself depends on the reality of '*sD*'-related Lines. Your theorem rests on two legs, and *both*, I fear, are rotten!

Nie. The next theorem is equivalent to Euc. I. 32. Do you wish to hear it?

Min. It is unnecessary: it follows easily from Th. 8.

And I need not ask you what practical test you provide for the meeting of two Lines, seeing that you have Euclid's 12th Axiom itself.

Nie. Proved as a Theorem.

Min. *Attempted* to be proved as a Theorem. I will now take a hasty general survey of your client's book.

The first point calling for remark is the arrangement. You begin by dragging the unfortunate beginner straight into the most difficult part of the subject. Your first chapter positively bristles with difficulties about 'direction.' Then comes a long chapter on circles, including some very

complicated figures, and a theory of tangents which depends upon moving lines and vanishing chords—all most
disheartening to a beginner. What do you suppose he
is likely to make of such a sentence as ' the direction of
the motion of the generating point of any curve is that
of the tangent to the curve at that point'? (p. 29.) Or
this again, 'it is also evident that, the circle being a
simple curve, there can be only *one* tangent to it at any
point'? (p. 29.) What *is* ' a simple curve'?

Nie. I do not know.

Min. Then comes a chapter of problems, and *then*—
when your pupil has succeeded in mastering thirty-four
pages of your book, and has become tolerably familiar
with tangents and segments; with diametral lines and
reëntrant angles, with ' oval forms' and 'forms semi-
convex, semi-concave,'—you at last confront him with
that abstruse and much dreaded theorem, Euc. I. 4!
True, he has the ' Asses' Bridge ' to help him in proving
it, that in its turn being proved, apparently, by properties
of the circle; but, even with all these assistances, it is an
arduous task !

Nie. You are hard on my client.

Min. Well, jesting apart, let me say in all seriousness
that I think it would require very great ingenuity to
make a worse arrangement of the subject of Geometry,
for purposes of teaching, than is to be found in this little
book.

I do not think it necessary to criticise the book throughout: but I will mention one or two passages which have
caught my eye in glancing through it.

Here, for instance, is something about ' Directives,'
which seem to be a curious kind of Loci—quite different
from Right Lines, I should say.

Nie. Oh no !   They are exactly the same thing !

Min. Well, I find, at p. 4, ' Directives are either diver-
gent or parallel ' : and again, at p. 11, ' Parallel Directives
cannot meet.'   Clearly, then, Directives can never by any
possibility *coincide* : but ordinary Right Lines occasionally
do so, do they not ?

Nie. It is a curious *lapsus pennae.*

Min. At p. 7, I observe an article headed ' The principle
of double conversion,' which I will quote entire.

Reads.

' If four magnitudes, *a, b, A, B,* are so related, that when
a is greater than *b, A* is greater than *B* ; and when *a* is
equal to *b, A* is equal to *B* : then, *conversely,* when *A* is
greater than *B, a* is greater than *b ;* and, when *A* is equal
to *B, a* is equal to *b.*

' The truth of this principle, which extends to every
kind of magnitude, is thus made evident :—If, when *A*
is greater than *B, a* is not greater than *b,* it must be either
less than or equal to *b.*   But it cannot be less ; for, if it
were, *A* should, by the antecedent part of the proposition,
be less than *B,* which is contrary to the supposition made.
Nor can it be equal to *b ;* for, in that case, *A* should be
equal to *B,* also contrary to supposition.   Since, therefore,
a is neither less than nor equal to *b,* it remains that it must
be greater than *b.*'

Now let *a* and *A* be variables and represent the ordinates to two curves, *mnr* and *MNR*, for the same abscissa; and let *b* and *B* be constants and represent their intercepts on the *Y*-axis; i.e. let $On = b$, and $ON = B$.

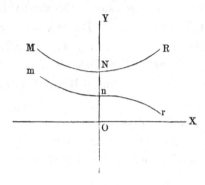

Does not this diagram fairly represent the *data* of the proposition? You see, when we take a negative abscissa, so as to make *a* greater than *b*, we are on the left-hand branch of the curve, and *A* is also greater than *B*; and again, when *a* is equal to *b*, we are crossing the *Y*-axis, where *A* is also equal to *B*.

Nie. It seems fair enough.

Min. But the conclusion does not follow? With a positive abscissa, *A* is greater than *B*, but *a* less than *b*.

Nie. We cannot deny it.

Min. What then do you suppose would be the effect on a simple-minded student who should wrestle with this terrible theorem, firm in the conviction that, being in a printed book, it must *somehow* be true?

Nie. (*gravely*) *Insomnia*, certainly; followed by acute *Cephalalgia*; and, in all probability, *Epistaxis*.

Min. Ah, those terrible names! Who would suppose

that a man could have all those three maladies, and survive? And yet the thing is possible!

Let me now read you a statement (at p. 112) about incommensurables :—

'When one of the magnitudes can be represented only by an interminable decimal, while the other is a finite whole number, or finite decimal, no finite common submultiple can exist; for, though a unit be selected in the last place of the whole number or finite decimal, yet the decimal represented by all the figures which follow the corresponding place in the interminable decimal, being less than that unit in that place and unknown in quantity, cannot be a common measure of the two magnitudes, and is only a remainder.'

Now can you lay your hand upon your heart and declare, on the word of an honest man, that you understand this sentence—beginning at the words ' yet the decimal '?

Nie. (vehemently) I cannot!

Min. Of the two reasons which are mentioned, to explain why it ' cannot be a common measure of the two magnitudes,' does the first—that it is ' less than that unit in that place '—carry conviction to your mind? And does the second—that it is ' unknown in magnitude ' ripen that conviction into certainty?

Nie. (wildly) Not in the least!

Min. Well, I will not 'slay the slain' any longer. You may consider Dr. Willock's book as rejected. And I think we may say that the whole theory of ' direction ' has collapsed under our examination.

Nie. I greatly fear so.

K 2

ACT III.

Scene I.

§ 1. The other Modern Rivals.

'But mice, and rats, and such small deer,
Have been Tom's food for seven long year.'

Min. I consider the question, as to whether Euclid's system and numeration should be abandoned or retained, to be now set at rest: the subject of Parallels being disposed of, no minor points of difference can possibly justify the abandonment of our old friend in favour of any Modern Rival. Still it will be worth while to examine the other writers, whose works you have brought with you, as they may furnish some valuable suggestions for the improvement of Euclid's Manual.

Nie. The other writers are Chauvenet, Loomis, Morell, Reynolds, and Wright.

Min. There are a few matters, as to which we may consider them all at once. How do they define a straight Line?

Nie. All but Mr. Reynolds define it as the shortest distance between two points, or more accurately, to use the words of Mr. Chauvenet, 'a line of which every portion is the shortest line between the points limiting that portion.

Min. We discussed that definition in M. Legendre's book. How does Mr. Reynolds define it?

Nie. Not at all.

Min. Very cautious. What of angles?

Nie. Some of them allow larger limits than Euclid does. Mr. Wright talks about 'angles of continuation' and 'angles of rotation.'

Min. Good for Trigonometry: not so suitable to early Geometry. How do they define Parallels?

Nie. As in Euclid, all of them.

Min. And which proposition of Tab. II. do they assume?

Nie. Playfair's, or else its equivalent, 'only one Line can be drawn, parallel to a given Line, through a given point outside it.'

Min. Now let us take them one by one.

ACT III.

§ 2.   Chauvenet.

'Where Washington hath left
His awful memory
A light for after times!'

Nie. I lay before you '*A Treatise on Elementary Geometry*,'
by W. Chauvenet, LL.D., Professor of Mathematics and
Astronomy in Washington University, published in 1876.

Min. I read in the Preface (p. 4) 'I have endeavoured
to set forth the elements with all the rigour and completeness demanded by the present state of the general science,
*without seriously departing from the established order of the
propositions.*' So there would be little difficulty, I fancy,
in introducing into Euclid's own Manual all the improve-
ments which Mr. Chauvenet can suggest.

P. 14. Pr. 1, and p. 18. Pr. v, taken together, tell us
that only one perpendicular can be drawn to a Line from a
point. And various additions, about obliques, are made in
subsequent propositions. All these may well be embodied
in a new proposition, which we might interpolate as Euc. I.
12. B.

P. 26. Pr. xv, asserts the equidistance of Parallels. This might be interpolated as Euc. I. 34. B.

Another new theorem, that angles whose sides are parallel, each to each, are equal (which I observe is a great favourite with the Modern Rivals), seems to me a rather clumsy and uninteresting extension of Euc. I. 29.

I see several propositions which might well be inserted as *exercises* on Euclid (*e. g.* Pr. xxxix, 'Every point in the bisector of an angle is equally distant from the sides'), but which are hardly of sufficient importance to be included as propositions: and others (*e. g.* Pr. xl, 'The bisectors of the three angles of a triangle meet in the same point) which seem to belong more properly to Euc. III or IV. I have no other remarks to make on this book, which seems well and clearly written.

ACT III.

SCENE I.

§ 3. LOOMIS.

'Like—but oh! how different!'

Nie. I lay before you '*Elements of Geometry*,' by ELIAS LOOMIS, LL.D., Professor of Natural Philosophy and Astronomy in Yale ˙College, a revised edition, 1876.

Min. I read in the Preface (p. 10) 'The present volume follows substantially the order of Blanchet's Legendre, while the form of the demonstrations is modeled after the more logical method of Euclid.' He has not, however, adopted the method of infinite series, which constitutes the crucial distinction between that writer and Euclid.

The propositions are pretty nearly in Euclid's order: with a few changes in order and numeration, the book would be a modernised Euclid, the only important differences being the adoption of Playfair's Axiom and the omission of the diagonals in Euc. II. I have no hostile criticisms to offer. Our American cousins set us an excellent example in the art of brief, and yet lucid, mathematical writing.

ACT III.

Scene I.

§ 4. Morell.

'Quis custodiet ipsos custodes?
Quis inspiciet ipsos inspectores?'

Nie. I lay before you '*Euclid Simplified, compiled from the most important French works, approved by the University of Paris and the Minister of Public Instruction,*' by Mr. J. R. Morell, formerly H. M. Inspector of Schools, published in 1875.

Min. What have you about Lines, to begin with?

Nie. Here is a definition. 'The place where two surfaces meet is called a Line.'

Min. Really! Let us take two touching spheres, for instance?

Nie. Ahem! We abandon the definition.

Min. Perhaps we shall be more fortunate with the definition of a *straight* Line.

Nie. It is 'an indefinite Line, which is the shortest between any two of its points.'

Min. An '*indefinite*' Line! What in the world do you mean? Is a curved Line more definite than a straight Line?

Nie. I don't know.

Min. Nor I. The rest of the sentence is slightly elliptical. Of course you mean 'the shortest which can be drawn'?

Nie. (*eagerly*) Yes, yes!

Min. Well, we have discussed that matter already. Go on.

Nie. Next we have an axiom, 'that from one point to another only one straight Line can be drawn, and that if two portions of a straight Line coincide, these Lines coincide throughout their whole extent.'

Min. You bewilder me. How can one portion of a straight Line coincide with another?

Nie. (*after a pause*) It can't, of course, *in situ* : but why not take up one portion and lay it on another?

Min. By all means, if you like. Let us take a certain straight Line, cut out an inch of it, and lay it along another inch of the Line. What follows?

Nie. Then 'these Lines coincide throughout their whole extent.'

Min. Do they indeed? And pray who *are* 'these Lines'? The two inches?

Nie. (*gloomily*) I suppose so.

Min. Then the axiom is simple tautology.

Nie. Well then, we mean the whole straight Line and—and—

Min. And what else? You can't talk of 'one straight Line' as 'these Lines,' you know.

Nie. We abandon the axiom.

Min. Better luck next time! Try another definition.

Nie. 'A broken Line is a Line composed of straight Lines.'

Min. But a *straight* Line also is 'a Line composed of straight Lines,' isn't it?

Nie. Well, we abandon the definition.

Min. This is quite a new process in our navigation. Instead of heaving the lead, we seem to be throwing overboard the whole of our cargo! Let us hear something about Angles.

Nie. 'The figure formed by two Lines that intersect is called an Angle.'

Min. What do you mean by 'figure'? Do you define it anywhere?

Nie. Yes. 'The name of figure is given to volumes, surfaces, and lines.'

Min. Under which category do you put 'Angle'?

Nie. I don't know.

Min. Anything new about the definition, or equality, of right angles?

Nie. No, except that we *prove* that all right angles are equal.

Min. That we have discussed already (see p. 56). Let us go on to Pairs of Lines, and your proof of Euc. I. 29, 32.

Niemand *reads.*

'Th. 19. Two Lines perpendicular to the same Line are parallel.'

Min. Do you mean 'separational'?

Nie. Yes.

Min. Have you defined 'parallel' anywhere?

Nie. (after a search) I can't find it.

Min. A careless omission. However, your assertion is that 't_r ℙ S.' *That* isn't always true. Suppose your two Lines were drawn from the same point?

Nie. We beg to correct the sentence. 'Two *different* Lines.'

Min. Very well. Then you assert 'st_r ℙ S' (Tab. I. 5). I grant it.

NIEMAND *reads.*

'Th. 20. Through a point situated outside a straight Line a parallel, and only one, can be drawn to that Line.'

Min. '*A* parallel,' I grant at once: it is 'sS real' (Tab. I. 9). But '*only one*'! That takes us into Table II. What axiom do you assume?

Nie. 'It may be admitted that only one parallel can be drawn to it.'

Min. That is 'sS unique'—a contranominal of Playfair's Axiom. Very well. We need not pursue the subject: all is easy after that. Now hand *me* the book, if you please: I wish to make a general survey of style, &c.

At p. 4 I read :—' Two theorems are reciprocal when the hypothesis and the conclusion of one are the conclusion and the hypothesis of the other.' (They are usually called 'converse'—the *technical,* not the *logical,* converse, as was mentioned some time ago (p. 46); but let that pass.)

' Thus the theorem—*if two angles are right angles, they are equal*—has for its reciprocal—*if two angles are equal, they are right angles.*'

(This, by the way, is a capital instance of the distinction between ' technical ' and ' logical.' Here the *technical* converse is wild nonsense, while the *logical* converse is of course as true as the theorem itself: it is ' *some cases of two angles being equal are cases of their being right.*')

' All propositions are direct, reciprocal, or contrary—all so closely connected that either of the two latter ' (I presume he means ' the latter two ') ' is a consequence of the other two.'

A ' consequence '! Can he mean a *logical* consequence ? Would he let us make a syllogism of the three, using the ' direct ' and ' reciprocal ' (for instance) as premisses, and the ' contrary ' as the conclusion?

However, let us first see what he means by a ' contrary ' proposition.

' It is a direct proposition to prove that all points in a circle enjoy a certain property, *e.g.* the same distance from the centre.'

(This notion of sentient points, by the way, is very charming. I like to think of all the points in a circle really feeling a placid satisfaction in the thought that they are equidistant from the centre ! They are infinite in number, and so can well afford to despise the arrogance of a point within, and to ignore the envious murmurs of a point without !)

' The contrary proposition shows that all points taken outside or inside the figure do not enjoy this property.'

So then this is his trio :—

    1. Direct. ' All X are Y.'

    2. Reciprocal. ' All Y are X.'

    3. Contrary. ' All not-X are not-Y.'

Here of course No. 2 and No. 3, being contranominals, are logically deducible from each other, No. 1 having no logical connection with either of them.

And yet he calls the three ' so closely connected that either of the two latter is a consequence of the other two'! Shade of Aldrich! Have we come to this? You say nothing, mein Herr?

Nie. I say that, if you grant what you call the 'premisses,' you cannot deny the conclusion.

Min. True. It reminds me of an answer given some years ago in the Schools at Oxford, when the Examiner asked for an example of a syllogism. After much patient thought, the candidate handed in

> ' All men are dogs;
> All dogs are men :
> *Therefore*, All men are dogs.'

This certainly has the *form* of a syllogism. Also it avoids, with marked success, the dangerous fallacy of ' four terms.' And it has the great merit of Mr. Morell's syllogism, that, if you grant the premisses, you cannot deny the conclusion. Nevertheless I feel bound to add that it was *not* commended by the Examiner.

Nie. I can well believe it.

Min. I proceed. ' The direct and the reciprocal proofs are generally the simpler, and do not require a fresh construction.' Why ' fresh'? The ' direct' comes *first*, apparently; so that, if it requires a construction at all, it *must* be a 'fresh' one.

Nie. Be not hypercritical.

Min. Well, it *is* rather 'small deer,' I confess: let us change the subject.

Here is a pretty proof in Th. 4.

> 'Then $m + o = m + x$.
> But $m = m$.
> Therefore $o = x$.'

Isn't that 'but $m = m$' a delightfully cautious parenthesis? Your client seems to be nearly as much at home in Algebra as in Logic, which is saying a great deal !

At p. 9, I read 'The base of an isosceles triangle is the unequal side.'

'*The unequal side*'! Is an equilateral triangle isosceles, or is it not? Answer, mein Herr !

Nie. Proceed.

Min. At p. 17, I read 'From one and the same point three equal straight Lines cannot be drawn to another straight Line; for if that were the case, *there would be on the same side of a perpendicular two equal obliques,* which is impossible.'

Kindly prove the italicised assertion on this diagram, in which I assume *FD, FC, FE,* to be equal lines, and

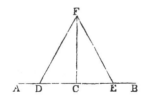

have made the *middle* one of the three a perpendicular to
the 'other straight Line.'

Nie. (*furiously*) I will not!

Min. Look at p. 36. 'A circumference is generally
described in language by one of its radii.' Let us hope
that the language is complimentary—at least if the circumference is within hearing! Can't you imagine the
radius gracefully rising to his feet, rubbing his lips with
his table-napkin? 'Gentlemen! The toast I have the
honour to propose is &c. &c. Gentlemen, I give you *the
Circumference*!' And then the chorus of excited Lines,
' For he's a jolly good felloe!'

Nie. (*rapturously*) Ha, ha! (*checking himself*) You are
insulting my client.

Min. Only filling in his suggestive outlines. Try p. 48.
'Th. 13. If two circumferences are interior,' &c. Can
your imagination, or mine, grasp the idea of two circumferences, each of them inside the other? No! *We*
are mere prosaic mortals: it is beyond us!

In p. 49 I see some strange remarks about ratios. First
look at Def. 44. 'When a magnitude is contained an
exact number of times in two magnitudes of its kind, it is
said to be their common measure.' (The wording is awkward,
and suggests the idea of their having only *one* 'common
measure'; but let that pass.) 'The ratio of two magnitudes
of the same kind is the number which would express the
measure of the first, if the second were taken as unity.'

' *The measure of the first* '! Do you understand that?
Is it a 'measure' such as you have just defined? or some
other kind?

Nie. Some other kind, I *think*. But there is a slight obscurity somewhere.

Min. Perhaps this next enunciation will clear it up. 'If two magnitudes of the same kind, A and B, are mutually commensurable' (by the way, 'mutually' is tautology), 'their ratio is a whole or fractional number, which is obtained by dividing the two numbers one by the other, and which expresses how many times these magnitudes contain their common measure M.' Do you understand *that*?

Nie. Well, no!

Min. Let us take an instance—£3 and 10s. A shilling is *a* common measure of these two sums : will you accept it as '*their* common measure'?

Nie. We will do it, provisionally.

Min. Now the number 'obtained by dividing the two numbers' (I presume you mean 'the two magnitudes') 'one by the other' is '6,' is it not?

Nie. It would seem so.

Min. Well, does this number 'express how many times these magnitudes contain their common measure,' viz. a shilling?

Nie. Hardly.

Min. Did you ever meet with any *one* number that could 'express' *two* distinct facts?

Nie. We would rather change the subject.

Min. Very well, though there is plenty more about it, and the obscurity deepens as you go on. We will 'vary the verse' with a little bit of classical criticism. Look at p. 81. 'Homologous, from the Greek ὅμοιος, like or similar, λόγος, word or reason.' Do you think this school-inspector ever heard of the great Church controversy, where all turned on the difference between ὅμος and ὁμοῖος?

Nie. (*uneasily*) I think not. But this is not a *mathematical* silp, you know.

Min. You are right. *Revenons à nos moutons.* Turn to p. 145, art. 65. 'To measure areas, it is usual to take a square as unity.' To me, who have always been accustomed to regard 'a square' as a concrete magnitude and 'unity' as a pure number, the assertion comes rather as a shock. But I acquit the author of any intentional roughness. Nothing could surpass the delicacy of the next few words :—'It has been already stated that surfaces are measured indirectly'! Lines, of course, may be measured anyhow : *they* have no sensibilities to wound : but there is an open-handedness—a breadth of feeling—about a surface, which tells of noble birth—'every (square) inch a King!' —and so we measure it with averted eyes, and whisper its area with bated breath!

Nie. Return to other muttons.

Min. Well, take p. 156. Here is a 'scholium' on a theorem about the area of a sector of a circle. The 'scholium' begins thus :—'If *a* is the number of degrees in the arc of a sector, we shall have to find the length of this arc ——.' I pause to ask 'If *β* were the number, should we have to find it *then*?'

Nie. (*solemnly*) We should!

Min. 'For the two Lines which are multiplied in all rules for the measuring of areas must be referred to the same linear unity.' *That*, I take it, is fairly obscure : but it is luminous when compared with the note which follows it. 'If the linear unit and angular unit are left arbitrary, any angle has for measure the ratio of the numbers of linear units contained in the arcs which the angle in question and the irregular unit intercept in any circumference described from their summit as common centres.' Is not that a useful note? 'The irregular unit'! Linear, or angular, I wonder? And then 'common centres'! How many centres does a circumference usually require? I will only trouble you with one more extract, as a *bonne bouche* to wind up with.

'Th. 9. (P. 126.) *Every convex closed Line* ABCD *enveloped by any other closed Line* PQRST *is less than it.*

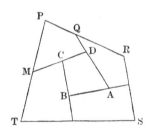

'All the infinite Lines *ABCD, PQRST,* &c.'—— by the way, these are curious instances of 'infinite Lines'?

Nie. (*hastily*) We mean 'infinite' in *number,* not in *length.*

Min. Well, you express yourself oddly, at any rate. '—— which enclose the plane surface *ABCD,* cannot be equal. For drawing the straight Line *MD,* which does not cut *ABCD, MD* will be less than *MPQD*; and adding to both members the part *MTSRQD*, the result will be *MDQRSTM* less than *MPQRSTM.*' Is that result proved?

Nie. No.

Min. Is it true?

Nie. Not necessarily so.

Min. Perhaps it is a *lapsus pennæ.* Try to amend it.

Nie. If we add to *MD* the part *MTSRQD*, we do get *MDQRSTM*, it is true: but, if we add it to *MPQD*, we get *QD* twice over; that is, we get *MPQRSTM* together with twice *QD*.

Min. How does that addition suit the rest of the proof?

Nie. It ruins it: all depends on our proving the perimeter *MDQRSTM* less than the perimeter *MPQRSTM*, which this method has failed to do—as of course all methods must, the thing not being capable of proof.

Min. Then the whole proposition breaks down absolutely?

Nie. We cannot deny it.

Min. Let us turn to the next author.

ACT III.

SCENE I.

§ 5. REYNOLDS.

'Though this be madness, yet there's method in 't.'

Nie. I lay before you '*Modern Methods in Elementary Geometry,*' by E. M. REYNOLDS, M.A., Mathematical Master in Clifton College, Modern Side; published in 1868.

Min. The first remark I have to make on it is, that the Definitions and Axioms are scattered through the book, instead of being placed together at the beginning, and that there is no index to them, so that the reader only comes on them by chance: it is quite impossible to refer to them.

Nie. I cannot defend the innovation.

Min. In Th. 1 (p. 3), I read 'the angles *CDA*, *CDB* are together equal to two right angles. *For they fill exactly the same space.*' Do you mean finite or infinite space? If 'finite,' we increase the angle by lengthening its sides: if 'infinite,' the idea is unsuited for elementary teaching. You had better abandon the idea of an angle 'filling space,' which is no improvement on Euclid's method.

P. 61. Th. II (of Book III) it is stated that parallelograms, on equal bases and between the same parallels,
'may always be placed so that their equal bases coincide,'
and it is clearly assumed that they will still be 'between
the same parallels.' And again, in p. 63, the altitude of a
parallelogram is defined as '*the* perpendicular distance of
the opposite side from the base,' clearly assuming that
there is only *one* such distance. In both these passages
the theorem is assumed 'Parallels are equidistant from each
other,' of which no proof has been given, though of course
it might have been easily deduced from Th. XVI (p. 19).

The theorems in Euc. II are here proved algebraically,
which I hold to be emphatically a change for the worse,
chiefly because it brings in the difficult subject of incom-
mensurable magnitudes, which should certainly be avoided
in a book meant for beginners.

I have little else to remark on in this book. Several of
the new theorems in it seem to me to be premature, e. g.
Th. XIX, &c. on 'Loci': but the sins of *omission* are more
serious. He actually leaves out Euc. I. 7, 17, 21 (2nd
part), 24, 25, 26 (2nd part), 48, and II. 1, 2, 3, 8, 9, 10, 12,
13. Moreover he separates Problems and Theorems, which
I hold to be a mistake. I will not trouble you with any
further remarks.

ACT III.

SCENE I.

§ 6. WRIGHT.

' Defects of execution unquestionably exist.'
WRIGHT, *Pref.* p. 10.

Nie. I lay before you ' *The Elements of Plane Geometry,*' by R. P. WRIGHT, Teacher of Mathematics in University College School, London; the second edition, 1871.

Min. Some of the changes in Euclid's method, made in this book, are defended in the Preface.

First, he claims credit for having more Axioms than Euclid, whom he blames for having demonstrated 'much that is obvious.' I need hardly pause to remind you that ' obviousness ' is not an invariable property : to a *perfect* intellect the whole of Euclid, to the end of Book XII, would be ' obvious ' as soon as the Definitions had been mastered : but Geometricians must write for *imperfect* intellects, and it cannot be settled on general principles where axioms should end and theorems begin. Let us look at a few of these new axioms. In p. viii of the Preface, I read 'with the conception of straightness in a line we naturally associate that of the utmost possible shortness of path between any two of its points; allow this to be assumed, &c.' This I consider a most objectionable axiom, obliging us, as it does, to contemplate the lengths of *curved* lines. This matter I have already discussed with M. Legendre (p. 55).

Secondly, for the host of new axioms with which we are threatened in the Preface, I have searched the book in vain: possibly I have overlooked some, as he never uses the heading 'Axiom,' but really I can only find *one* new one, at p. 5. 'Every angle has one, and only one bisector,' which is hardly worth stating. Perhaps the writer means that his proofs are not so full as those in Euclid, but take more for granted. I do not think this any improvement in a book meant for beginners.

Another change, claimed in the Preface as an improvement, is the more constant use of superposition. I have considered that point already (p. 46) and have come to the conclusion that Euclid's method of constructing a new figure has all the advantages, without the obscurity, of the method of superposition.

I see little to remark on in the general style of the book. At p. 21 I read 'the straight line *AI* satisfies the four following conditions: it passes through the vertex *A*, through the middle point *I* of the base, is a perpendicular on that base, and is the bisector of the vertical angle. Now, two of these four conditions suffice to determine the straight line *AI*, . . . Hence a straight line fulfilling any two of these four conditions necessarily fulfils the other two.' All this is strangely inaccurate : the fourth condition is sufficient by itself to determine the line *AI*.

At p. 40 I notice the startling announcement that 'the simplest of all polygons is the *triangle*'! This is surely a new use for 'many'? I wonder if the writer is prepared to accept the statement that '*many* people have swum across the Bosphorus' on the strength of Byron's

> ' As once (a feat on which ourselves we prided)
> Leander, Mr. Ekenhead, and I did.'

As a specimen of the wordy and unscientific style of the writer, take the following :—

'*From any point O, one, and only one, perpendicular can be drawn to a given straight line AB.*

' Let *O'* be the point on which *O* would fall if, the paper being folded along *AB*, the upper portion of the figure were turned down upon the lower portion. If from the points *O, O'* straight lines be drawn to any point whatever *I* on the line *AB*, the adjacent angles *OIB, O'IB* will be equal ; for folding the paper again along *AB* and turning the upper portion down upon the lower, *O* falls on *O'*, *I* remains fixed, and the angle *OIB* exactly coincides with *O'IB*. Now in order that the line *OI* may be perpendicular to *AB*, or, in other words, that *OIB* may be a right angle, the sum of the two adjacent angles OIB, $O'IB$ must be equal to two right angles, and consequently their sides IO, IO' in the same straight line. But since we can always draw one, and only one, straight line between two points O and O', it follows that from a point O we can always draw one, and only one, perpendicular to the line AB.'

Do you think you could make a more awkward or more obscure proof of this almost axiomatic theorem ?

Nie. (cautiously) I would not undertake it.

Min. All that about folding and re-folding the paper is more like a child's book of puzzles than a scientific treatise. I should be very sorry to be the school-boy who is expected to *learn* this precious demonstration ! In such a case, I could not better express my feelings than by quoting three words of this very theorem :—'*I remains fixed* ' !

In conclusion, I may say as to all five of these authors, that they do not seem to me to contain any desirable novelty which could not easily be introduced into an amended edition of Euclid.

Nie. It is a position I cannot dispute.

ACT III.

Scene II.

§ 1. Syllabus of the Association for the Improvement of Geometrical Teaching. 1878.

'Nos numerus sumus.'

Nie. The last book to be examined is Mr. Wilson's new Manual, founded on the Syllabus of the Geometrical Association.

Min. We had better begin by examining the Syllabus itself. I own that I could have wished to do this in the presence of some member of the Committee, who might have supplied a few details for what is at present little more than a skeleton, but that I fear is out of the question.

Nie. Nay, you shall not have far to seek. *I* am a member of the Committee.

Min. (astonished) You! A German professor! No such member is included in the final list of the Committee, which a friend showed me the other day.

Nie. The final list, was it ?   Well, ask your friend whether, since the drawing up of that list, any addition has been made : he will say ' Nobody has been added.'

Min. Quite so.

Nie. You do not understand.   *Nobody—Niemand*—see you not ?

Min. What ?   You mean—

Nie. (*solemnly*) I do, my friend.   *I* have been added to it !

Min. (*bowing*) The Committee are highly honoured, I am sure.

Nie. So they ought to be, considering that I am a more distinguished mathematician than Newton himself, and that *my* Manual is better known than Euclid's !   Excuse my self-glorification, but any moralist will tell you that I—I alone among men—*ought* to praise myself.

Min. (*thoughtfully*) True, true.   But all this is word-juggling—a most misleading analogy.   However, as you now appear in a new character, you must at least have a new name !

Nie. (*proudly*) Call me *Nostradamus !*

[*Even as he utters the mystic name, the air grows dense around him, and gradually crystallizes into living forms. Enter a phantasmic procession, grouped about a banner, on which is emblazoned in letters of gold the title '* ASSOCIATION FOR THE IMPROVEMENT OF THINGS IN GENERAL.' *Foremost in the line marches* NERO, *carrying his unfinished ' Scheme for the Amelioration of Rome'; while among the crowd which follow him may be noticed*—GUY FAWKES, *President of the ' Association for raising the position of Members of Parlia-*

ment'—THE MARCHIONESS DE BRINVILLIERS, *Foundress of the ' Association for the Amendment of the Digestive Faculty '* —*and* THE REV. F. GUSTRELL (*the being who cut down Shakspeare's mulberry-tree*), *leader of the 'Association for the Refinement of Literary Taste.' Afterwards enter, on the other side, Sir Isaac Newton's little dog 'DIAMOND,' carrying in his mouth a half-burnt roll of manuscript. He pointedly avoids the procession and the banner, and marches past alone, serene in the consciousness that he, single-pawed, conceived and carried out his great ' Scheme for the Advancement of Mathematical Research,' without the aid of any Association whatever.*]

Min. Nostra, the plural of *nostrum*, 'a quack remedy'; and *damus*, 'we give.'   It is a suggestive name.

Nos. And, trust me, it is a suggestive book that I now lay before you.   ' *Syllabus*—'.

Min. (*interrupting*) You mean ' *a* Syllabus ', or ' *the* Syllabus '?

Nos. No, no!   In this railroad-age, we have no time for superfluous words!   ' *Syllabus of Plane Geometry, prepared by the Association for the Improvement of Geometrical Teaching.*'   Fourth Edition, 1877.

Min. How do you define a Right Line?

NOSTRADAMUS *reads.*

P. 7. Def. 5. 'A straight line is such that any part will, however placed, lie wholly on any other part, if its extremities are made to fall on that other part.'

Min. That looks more like a *property* of a Right Line than its *essence.*   Euclid makes an axiom of that property. Of course you omit his axiom?

Nos. No. We have the axiom (p. 10, Ax. 2) 'Two straight lines that have two points in common lie wholly in the same straight line.'

Min. Well! That is certainly the strangest axiom I ever heard of! The idea of asserting, as an axiom, that Right Lines answer to their definition!

Nos. (*bashfully*) Well, you see there were several of us at work drawing up this Syllabus: and we've got it a little mixed: we don't quite know which are definitions and which are axioms.

Min. So it appears: not that it matters much: the practical test is the only thing of importance. Do you adopt Euc. I. 14?

Nos. Yes.

Min. Then we may go on to the next subject. Be good enough to define 'Angle.'

NOSTRADAMUS *reads.*

P. 8. Def. 11. 'When two straight lines are drawn from the same point, they are said to contain, or to make with each other, a *plane angle.*'

Min. Humph! You are very particular about drawing them *from* a point. Suppose they were drawn *to* the same point, what would they make then?

Nos. An angle, undoubtedly.

Min. Then why omit that case? However, it matters little. You say '*a* plane angle,' I observe. You limit an angle, then, to a magnitude less than the sum of two right angles.

Nos. No, I can't say we do. A little further down we

assert that ' *two* angles are formed by two straight lines drawn from a point.'

Min. Why, these are like Falstaff's ' rogues in buckram suits ' ! Are there more coming ?

Nos. No, we do not go beyond the sum of four right angles. These two we call *conjugate* angles. ' The greater of the two is called the *major conjugate*, and the smaller the *minor conjugate*, angle.'

Min. These definitions are wondrous ! This is the first time I ever heard ' major ' and ' minor ' defined. One feels inclined to say, like that Judge in the story, when a certain barrister, talking against time, insisted on quoting authorities for the most elementary principles of law, ' Really, brother, there are *some* things the Court may be assumed to know ! ' Any more definitions ?

Nos. We define ' a straight angle.'

Min. That I have discussed already (see p. 75).

Nos. But *this*, I think, is new :—

Reads.

P. 9. Def. 12. ' When three straight lines are drawn from a point, if one of them be regarded as lying between the other two, the angles which this one (the mean) makes with the other two (the extremes) are said to be *adjacent* angles.'

Min. That is new indeed. Let us try a figure :—

Now let us regard OA 'as lying between the other two.'
Which are '*the* angles which it makes with the other two'?
For this line OA (which you rightly call ' the mean '—
lying is always mean) makes, be pleased to observe, *four*
angles altogether—two with OB, and two with OC.

Nos. I cannot answer your question. You confuse me.

Min. I need not have troubled you. I see that I can
obtain an answer from the Syllabus itself. It says (at
the end of Def. 11) ' when *the angle contained by two lines*
is spoken of without qualification, the *minor conjugate* angle
is to be understood.' Here we have a case in point, as
these angles are spoken of ' without qualification.' So that
the angles alluded to are both of them ' minor conjugate '
angles, and lie on the same side of OA. And these we are
told to call ' adjacent ' angles !

How do you define a Right Angle ?

Nos. As in Euclid.

Min. Let me hear it, if you please. You know Euclid
has no major or minor conjugate angles.

Nostradamus *reads*.

P. 9. Def. 14. ' When one straight line stands upon
another straight line and makes the adjacent angles equal,
each of the angles is called a *right angle*.'

Min. Allow me to present you with a figure, as I see the
Syllabus does not supply one.

A
B
C

Here AB ' stands upon ' BC and makes the adjacent angles
equal. How do you like these ' right angles '?

Nos. Not at all.

Min. These same ' conjugate angles ' will get you into many difficulties.

Have you Euclid's axiom ' all right angles are equal ' ?

Nos. Yes, only we propose to prove it as a theorem.

Min. I have no objection to that : nor do I think that your treatment of angles, as a whole, is actually *illogical.* What I chiefly object to is the general ' slipshoddity ' (if I may coin a word) of the language of your Syllabus.

Does your proof of Euc. I. 32 differ from his ?

Nos. No, except that we propose Playfair's axiom, ' two straight Lines that intersect one another cannot both be parallel to the same straight Line,' as a substitute for Euc. Ax. 12.

Min. Is this your only test for the meeting of two Lines, or do you provide any other ?

Nos. This is the only one.

Min. But there are cases where this is of no use.   For instance, if you wish to make a triangle, having, as *data*, a side and the two adjacent angles.   Have you such a problem ?

Nos. Yes, it is Pr. 10, at p. 19.

Min. And how do you prove that the Lines will meet ?

Nos. (*smiling*) We *don't* prove it : that is the reader's business : we only provide enunciations.

Min. You are like the *gourmand* who would eat so many oysters at supper that at last his friend could not help saying ' They are sure to disagree with you in the night.' ' That is *their* affair,' the other gaily replied.   ' *I* shall be asleep ! '

Your Syllabus has the same hiatus as the other writers
who have rejected Euclid's 12th Axiom. If you will not
have it as an axiom, you ought to prove
it as a theorem. Your treatise is incom-
plete without it.

| 13—15 |
| 4, 5 |
| 26 a |
| 6 |
| 16 |
| 18—24 |
| 8 |
| 25 |
| 26 β |
| 17 |

The Theorems contained in the first 26
Propositions of Euclid are thus rearranged
in the Syllabus. The only advantage that
I can see in the new arrangement is that
it places first the three which relate to
Lines, thus getting all those which relate
to Triangles into a consecutive series. All
the other changes seem to be for the worse,
and specially the separation of Theorems
from their converses, *e.g.* Props. 5, 6, and
24, 25.

The third part of Prop. 29 is put after Prop. 32 : and
Props. 33, 34 are transposed. I can see no reason for
either change.

Prop. 47 is put next before Prop. 12 in Book II. This
would be a good arrangement (if it were ever proved to
be worth while to abandon Euclid's order), as the Theorems
are so similar; and the placing Prop. 48 next after II. 13
is a necessary result.

In Book II, Props. 9, 10 are placed after Props. 12, 13.
I see no reason for it.

It does not appear to me that the new arrangements,
for the sake of which it is proposed to abandon the numeration of Euclid, have anything worth mentioning to offer
as an advantage.

I will now go through a few pages of 'this many-headed monster,' and make some general remarks on its *style*.

P. 4. 'A *Theorem* is the formal statement of a proposition that may be demonstrated from known propositions. These known propositions may themselves be Theorems or Axioms.'

This is a truly delightful jumble. Clearly, 'a proposition that may be demonstrated from known propositions' is itself a Theorem. Hence a Theorem is 'the formal statement' *of* a Theorem. The question now arises—of itself, or of some other Theorem? That a Theorem should be 'the formal statement' of *itself*, has a comfortable domestic sound, something like 'every man his own washerwoman,' but at the same time it involves a fearful metaphysical subtlety. That one Theorem should be 'the formal statement' of *another* Theorem, is, I think, degrading to the former, unless the second will consent to act on the 'claw me, claw thee' principle, and to be 'the formal statement' of the first.

Nos. (*wildly*) Say no more! My brain reels!

Min. I spare you. Let us go on to p. 5, where I find the following :—

'*Rule of Conversion.* If of the hypotheses of a group of demonstrated Theorems it can be said that one must be true, and of the conclusions that no two can be true at the same time, then the converse of every Theorem of the group will necessarily be true.'

Let us take an instance :—

$$\text{If } 5 > 4, \text{ then } 5 > 3.$$
$$\text{If } 5 < 2, \text{ then } 5 < 3.$$

M 2

Those will do for ' demonstrated Theorems,' I suppose?

Nos. I suppose so.

Min. And the ' hypothesis' of the first ' must be true,' simply because it *is* true.

Nos. It would seem so.

Min. And it is quite clear that ' of the conclusions no two can be true at the same time,' for they contradict each other.

Nos. Clearly.

Min. Then it ought to follow that ' the converse of every Theorem of the group will necessarily be true.' Take the converse of the second, i. e.

$$\text{If } 5 < 3, \text{ then } 5 < 2.$$

Is this 'necessarily true'? Is every thing which is less than 3 necessarily less than 2?

Nos. Certainly not. I think you have misinterpreted the phrase ' it can be said that one must be true,' when used of the hypotheses. It does not mean ' it can be said, from a knowledge of the subject-matter of some *one* hypothesis, that it *is*, and therefore must be, true,' but ' it can be said, from a knowledge of the mutual logical relation of *all* the hypotheses, as a question of *form* alone, and without any knowledge of their subject-matter, that one must be true, though we do not know which it is.'

Min. Your power of uttering long sentences is one that does equal honour to your head and your—lungs. And most sincerely do I pity the unfortunate learner who has to make out all that for himself! Let us proceed.

P. 9. Def. 13. ' The *bisector* of an angle is the straight Line that divides it into two equal angles.'

This assumes that 'an angle has one and only one bisector,' which appears as Ax. 4, at the foot of p. 10.

P. 10. Def. 21. 'The opposite angles made by two straight Lines that intersect &c.'

This seems to imply that 'two Lines that intersect' always *do* make 'opposite angles.'

Nos. Surely they do?

Min. By no means. Look at p. 12, Def. 32, where, in speaking of a triangle, you say 'the intersection of the other two sides is called the vertex.'

Nos. A slip, I confess.

Min. One of many.

P. 12, Def. 31. 'All other triangles are called acute-angled triangles.' What? If a triangle had two right angles, for instance?

Nos. But there *is* no such triangle.

Min. *That* is a point you do not prove till we come to Th. 18, Cor. 1, two pages further on. The same remark applies to your Def. 33, in the same page. 'The side . . . which is opposite to *the* right angle,' where you clearly assume that it cannot have more than one.

P. 12, Def. 32. 'When two of the sides have been mentioned, the remaining side is often called the base.' Well, but how if two of the sides have *not* been mentioned?

Nos. In that case we do not use the word.

Min. Do you not? Turn to p. 22, Th. 2, Cor. 1, 'Triangles on the same or equal bases and of equal altitude are equal.'

Nos. We abandon the point.

Min. You had better abandon the definition.

P. 12, Def. 34. Is not 'identically equal' tautology? Things that are 'identical' must surely be 'equal' also. Again, '*every part* of one being equal,' &c.   What do you mean by 'every part' of a rectilineal figure?

Nos. Its sides and angles, of course.

Min. Then what do you mean by *Ax (b)* in p. 3. 'The whole is equal to the sum of its parts'?   This time, I think I need not 'pause for a reply'!

P. 15, Def. 38. 'When a straight Line intersects two other straight Lines it makes with them eight angles etc.'

Let us count the angles at *G*.   They are, the 'major' and 'minor' angles which bear the name *EGA*; do. for *EGB*; do. for *AGH*; and do. for *BGH*.   That is, eight angles at *G* alone.   There are sixteen altogether.

P. 17, Th. 30. 'If a quadrilateral has two opposite sides equal and parallel, it is a parallelogram.'

This re-asserts part of its own data.

P. 17, Th. 31. 'Straight Lines that are equal and parallel have equal projections on any other straight Line; conversely, parallel straight Lines that have equal projections on another straight Line are equal.'

The first clause omits the case of Lines that are equal and in one and the same straight Line. The second clause

is not true: if the parallel Lines are at right angles to the other Line, their projections are equal, both being zero, whether the Lines are equal or not.

P. 18, Th. 32. 'If there are three parallel straight Lines, and the intercepts made by them on any straight Line that cuts them are equal, then etc.'

The subject of this proposition is inconceivable: there are *three* intercepts, and by no possibility can these three be equal.

P. 25, Prob. 5. 'To construct a rectilineal figure equal to a given rectilineal figure and having the number of its sides one less than that of the given figure.'

May I ask you to furnish me with the solution of this problem, taking, as your 'given rectilineal figure,' a triangle?

Nos. (indignantly) I decline to attempt it!

Min. I will now sum up the conclusions I have come to with respect to your Syllabus.

In the subjects of Lines, Angles, and Parallels, the changes you propose are as follows :—

You give a very unsatisfactory definition of a 'Right Line,' and then most illogically re-state it as an axiom.

You extend the definition of Angle—a most disastrous innovation.

Your definition of ' Right Angle ' is a failure.

You substitute Playfair's axiom for Euclid's 12th.

All these things are very poor compensation indeed for the vital changes you propose — the separation of Problems and Theorems, and the abandonment of Euclid's order and numeration. Restore the Problems (which are also Theorems) to their proper places, keep to Euclid's numbering (interpolating your new Propositions where you please), and your Syllabus may yet prove to be a valuable addition to the literature of Elementary Geometry.

ACT III.

Scene II.

---

§ 2. Wilson's 'Syllabus'-Manual.

'No followers allowed.'
TIMES' ADVERTISEMENT-SHEET, *passim.*

Nie. I lay before you '*Elementary Geometry, following the Syllabus prepared by the Geometrical Association,*' by J. M. WILSON, M.A., 1878.'

Min. In what respects is this book a 'Rival' of Euclid?

Nie. Well, it separates Problems from Theorems ——

Min. Already discussed (see p. 18).

Nie. It adopts Playfair's Axiom ——

Min. Discussed (see p. 42).

Nie. It abandons diagonals in Book II ——

Min. Discussed (see p. 49).

Nie. And it adopts a new sequence and numeration.

Min. That, of course, prevents us from taking it as merely a new edition of Euclid. It will need very strong evidence indeed to justify its claim to set aside the sequence and numeration of our old friend. We must now examine the book *seriatim.* When we come to matters that have been already condemned, either in Mr. Wilson's book, or in the 'Syllabus,' I shall simply note the fact. We need have no new discussion, except as to new matter.

Nie. Quite so.

Min. In the 'Introduction,' at p. 2, I read '*A Theorem is the formal statement of a Proposition*,' &c. Discussed at p. 163.

At p. 3 we have the 'Rule of Conversion,' which I have already endeavoured to understand (see p. 163).

At p. 6 is a really remarkable assertion. '*Every Theorem may be shewn to be a means of indirectly measuring some magnitude.*' Kindly illustrate this on Euc. I. 14.

Nie. (*hastily*) Oh, if you pick out one single accidental excep ——

Min. Well, then, take 16, if you like: or 17, or 18 ——

Nie. Enough, enough!

Min. (*raising his voice*)—or 19, or 20, or 21, or 24, or 25, or 27, or 28, or 30!

Nie. We abandon '*every.*'

Min. Good. At p. 8 we have the definitions of '*major conjugate*' and '*minor conjugate*' (discussed at p. 159).

At p. 9 is our old friend the '*straight angle*' (see p. 75).

In the same page we have that wonderful triad of Lines, one of which is '*regarded as lying between the other two*' (see p. 159).

And also the extraordinary result that follows when one straight Line '*stands upon another*' (see p. 160).

At p. 27, Theorem 14, is a new proof of Euc. I. 24, apparently an amended version of Mr. Wilson's five-case proof, which I discussed at p. 111. He has now reduced it to three cases, but I still think the '*bisector of the angle*' a superfluity.

At p. 37 we have those curious specimens of '*Theorems of Equality*,' which I discussed at p. 112.

At p. 53 is the Theorem which asserts, in its conclusion, part of its own *data* (see p. 166).

At p. 54 we are told that '*parallel lines, which have equal projections on another line, are equal*' (see p. 167).

At p. 55 we have the inconceivable triad of '*equal intercepts*' made by a Line cutting three Parallels (see p. 167).

At p. 161 I am surprised to see him fall into a trap in which I have often seen unwary students caught, while trying to say Euc. III. 30 ('To bisect a given arc'). After proving two chords equal, they at once conclude that *certain* arcs, cut off by them, are equal; forgetting to prove that the arcs in question are both *minor* arcs.

But I must go no further: I have already wandered beyond the limits of Euc. I, II. The one great merit of this book ——

Nie. You have mentioned all the *faults*, then?

Min. By no means. You are too impatient. The one great merit, as I was saying, of Mr. Wilson's new book (and a most blessed change it is!) is that it ignores the whole theory of 'direction.' That he has finally abandoned that night-mare of Elementary Geometry, I dare not hope: so all I have said about it had better stand, lest

in some future fit of inspiration he should bring out a
yet more agonising version of it.

But it has the usual hiatus of a system which replaces
Euclid's Axiom by Playfair's: it provides no means of
proving that particular finite Lines will meet if produced.
(This I have discussed at p. 161.)

Its proposed changes in the *sequence* of Euclid I have
discussed at p. 162.

It has a few other faults, which I have already discussed
in Mr. Wilson's own book, and a few peculiar to the
Syllabus; but I spare you such minute criticisms.

But what I have now to ask you is simply this. What
possible pretext have you left for suggesting that Euclid's
Manual, and specially his sequence and numeration, should
be abandoned in favour of this far from satisfactory infant?

Nie. There are some new Theorems ——

Min. Those constitute no reason: you might easily interpolate them.

Nie. I fear there are no other grounds to urge. But
I should like to consult the *doppelgänger* of the Association before I throw up my brief.

Min. By all means.

[*For a minute or two there is heard a rustling and a
whispering, as of ghosts. Then* NIEMAND *speaks again.*]

Nie. They think that, considering that this book is
but just published, and that it is definitely put forward
as *the* Manual to supersede Euclid, it ought to be examined more in detail, with reference to what is *new* in

it—that is, new proofs of Euclid's propositions, and new propositions.

Min. (with a weary sigh) Very well. It will perhaps be more satisfactory to do this, if only to ascertain exactly how much this new Manual contains that is really new and really worthy of adoption. But I shall limit my examination to the subject-matter of Euc. I, II.

Nie. That is all we ask.

Min. We begin, then, at p. 12.

Theorem 1. '*All right angles are equal.*' This is proved by their being halves of a 'straight angle,' a phrase which I have already criticised. There is a rather important omission in the proof, no distinction being drawn between the 'straight angle' on one side of a Line, and the other (of course named by the same letters) which lies on the other side and completes the four right angles. This Theorem, if proved without 'straight angles,' might be worth adding to a new edition of Euclid.

Th. 2 (p. 13) is Euc. I. 13, proved as in Euclid.

Th. 3 (p. 14) is Euc. I. 14, where, unfortunately, a new proof is attempted, which involves a fallacy. It is deduced from an 'Observation' in p. 9, that '*a straight line makes with its continuation at any point an angle of two right angles,*' which deduction can be effected only by the process of converting a universal affirmative '*simpliciter*' instead of '*per accidens.*'

Th. 4 (p. 14) is Euc. I. 15, proved as in Euclid.

At p. 17 I find a 'Question.' '*State the fact that " all geese have two legs " in the form of a theorem.*' This I would not mind attempting; but, when I read the additional

request, to ' *write down its converse theorem,*' it is so power-
fully borne in upon me that the writer of the Question
is probably himself a biped, that I feel I must, however
reluctantly, decline the task.

Th. 5 (p. 18) is Euc. I. 4, proved as in Euclid.

Th. 6 (p. 20) is Euc. I. 5, proved by supposing the
vertical angle to be bisected, thus introducing a 'hypothetical construction' (see p. 19).

Th. 7 (p. 21) is Euc. I. 26 (1st part), proved by superposition. Euclid's proof, by making a new triangle, is
quite as good, I think. The areas are here proved to be
equal, a point omitted by Euclid : I think it a desirable
addition to the theorem.

Th. 8 (p. 22) is Euc. I. 6, proved by reversing the
triangle and then placing it *on itself* (or on the trace it
has left behind), a most objectionable method (see p. 46).

Theorems 9 to 13 (pp. 22 to 26) are Euc. I. 16, 18, 19,
20, 21, with Euclid's proofs.

Th. 14 (p. 27) is Euc. I. 24, proved by supposing an
angle to be bisected : another 'hypothetical construction.'

Th. 15 (p. 28) is Euc. I. 8, for which two proofs are
offered :—one by Euc. I. 24 (which seems to be reversing
the natural order)—the other by an application of Euc.
I. 5, a method involving *three* cases, of which only one
is given. All this is to save the introduction of Euc.
I. 7, a theorem which *I* think should by no means be
omitted. (See p. 194.) Here, as in Th. 7, the equality
of the areas is, I think, a desirable addition to Euclid's
theorem.

Th. 16 (p. 29) is Euc. I. 25, with old proof.

Th. 17 (p. 30) is Euc. I. 26 (2nd part) proved by super-position instead of Euclid's method (which I prefer) of constructing a new triangle.

Th. 18 (p. 32) is Euc. I. 17, with old proof.

Th. 19 (p. 33) is new. '*Of all the straight lines that can be drawn from a given point to meet a given straight line, the perpendicular is the shortest; and of the others, those making equal angles with the perpendicular are equal; and that which makes a greater angle with the perpendicular is greater than that which makes a less.*' This I think deserves to be interpolated.

Th. 20 (p. 34) is new. '*If two triangles have two sides of the one equal to two sides of the other, each to each, and the angles opposite to two equal sides equal, the angles oppo-site to the other two equal sides are either equal or supple-mentary, and in the former case the triangles are equal in all respects.*' I do not think it worth while to trouble a beginner with this rather obscure theorem, which is of no practical use till he enters on Trigonometry.

Th. 21 (p. 43) is Euc. I. 27 : old proof.

Th. 22 (p. 44) is Euc. I. 29 (1st part), proved by Euc. I. 27 and Playfair's Axiom (see p. 42).

Th. 23 (p. 45) is new. '*If a straight line intersects two other straight lines and makes either a pair of alternate angles equal, or a pair of corresponding angles equal, or a pair of interior angles on the same side supplementary; then, in each case, the two pairs of alternate angles are equal, and the four pairs of corresponding angles are equal, and the two pairs of interior angles on the same side are supplementary.*' This most formidable enunciation melts

down into the mildest proportions when superfluities are
omitted, and only so much of it proved as is really necessary to include the whole. Euclid proves all that is
valuable in it in the course of I. 29, and I do not see
any sufficient reason for stating and proving it as a separate Theorem.

Th. 23, Cor. (p. 46) is the rest of Euc. I. 29 : old proof.

Th. 24 (p. 46) is Euc. I. 30, proved as a contranominal
of Playfair's Axiom.

Th. 25, 26, and Cor. (pp. 47, 48) are Euc. I. 32 and
Corollaries : old proof.

Th. 27, 1st part (p. 50), is a needless repetition of part
of the Corollary to Th. 23.

Th. 27, 2nd part (p. 50), is part of Euc. I. 34: old proof.

Th. 28 (p. 51) is the rest of Euc. I. 34, proved as in
Euclid.

Th. 29 (p. 52) is new. '*If two parallelograms have two
adjacent sides of the one respectively equal to two adjacent
sides of the other, and likewise an angle of the one equal
to an angle of the other ; the parallelograms are identically
equal.*' This might be a useful *exercise* to set ; but really
it does not seem of sufficient importance to be selected
for a Manual.

Th. 30 (p. 53) is Euc. I. 33 : old proof.

Th. 31 (p. 54) is new. '*Straight lines which are equal
and parallel have equal projections on any other straight
line ; conversely, parallel straight lines which have equal
projections on another straight line are equal ; and equal
straight lines, which have equal projections on another straight
line, are equally inclined to that line.*' The first and third

clauses might be interpolated, though I think their value doubtful. The second is false. (See p. 167.)

Th. 32 (p. 55) is new. '*If there are three parallel straight lines, and the intercepts made by them on any straight line that cuts them are equal, then the intercepts on any other straight line that cuts them are equal.*' This is awkwardly worded (in fact, as it stands, its subject, as I pointed out in p. 167, is inconceivable), and does not seem at all worth stating as a Theorem.

At p. 57 I see an 'Exercise' (No. 5). '*Shew that the angles of an equiangular triangle are equal to two-thirds of a right angle.*' In this attempt I feel sure I should fail. In early life I was taught to believe them equal to *two right angles*—an antiquated prejudice, no doubt; but it is difficult to eradicate these childish instincts.

Problem 1 (p. 61) is Euc. I. 9: old proof. It provides no means of finding a radius 'greater than half *AB*,' which would seem to require the previous bisection of *AB*. Thus the proof involves the fallacy '*Petitio Principii.*'

Pr. 2 (p. 62) is Euc. I. 11, proved nearly as in Euclid.

Pr. 3 (p. 62) is Euc. I. 12, proved nearly as in Euclid. It omits to say how a 'sufficient radius' can be secured, a point not neglected by Euclid.

Pr. 4 (p. 63) is Euc. I. 10, proved nearly as in Euclid. This also, like Pr. 1, involves the fallacy '*Petitio Principii.*'

Pr. 5 (p. 64) is Euc. I. 32, proved nearly as in Euclid, but claims to use compasses to transfer distances, a Postulate which Euclid has (properly, I think) treated as a Problem. (See p. 186.)

Pr. 6, 7 (pp. 65, 66) are Euc. I. 23 31 : old proofs.

Problems 8 to 11 (pp. 66 to 69) are new. Their object
is to construct triangles with various *data* : viz. *A, B,* and
c ; *A, B,* and *a* ; *a, b,* and *C* ; *a, b,* and *A.* They are
good exercises, I think, but hardly worth interpolating as
Theorems. The first of them is remarkable as one of the
instances where Mr. Wilson assumes Euc. Ax. 12, without
giving, or even suggesting, any proof. If he intends to
assume it as an *Axiom,* he makes Playfair's Axiom superfluous. No Manual ought to assume *both* of them.

Theorem 1 (p. 82) is Euc. I. 35, proved as in Euclid, but
incompletely, as it only treats of one out of three possible
cases.

Th. 2 (p. 83) is new. '*The area of a triangle is half the
area of a rectangle whose base and altitude are equal to those
of the triangle.*' This is merely a particular case of Euc. I.
41, and may fairly be reserved till we enter on Trigonometry, where it first begins to have any practical value.

Th. 2, Cor. 1 (p. 84) is Euc. I. 37, 38 : old proofs.

Th. 2, Cor. 2 (p. 84) is new. '*Equal triangles on the
same or equal bases have equal altitudes.*' No proof is
offered. It is an easy deduction, of questionable value.

Th. 2, Cor. 3 (p. 84) is Euc. I. 39, 40. No proof given.

Th. 3 (p. 84) is new. '*The area of a trapezium* [by
which Mr. Wilson means '*a quadrilateral that has only one
pair of opposite sides parallel*'] *is equal to the area of a
rectangle whose base is half the sum of the two parallel sides,
and whose altitude is the perpendicular distance between them.*'
I have no hesitation in pronouncing this to be a mere
' fancy' proposition, of no practical value whatever.

Th. 4 (p. 86) is Euc. I 43 : old proof.

Th. 5 (p. 87) is Euc. II. 1 : old proof.

Th. 6, 7, 8 (p. 88, &c.) are Euc. II. 4, 7, 5. The sequence of Euc. II. 5, and its Corollary, is here inverted. Also the diagonals are omitted, and nearly every detail is left unproved, thus attaining a charming brevity—of *appearance.*

Th. 9 (p. 91) is Euc. I 47 : old proof.

Th. 10, 11 (pp. 94, 95) are Euc. 12, 13 : old proof.

Th. 12 (p. 95) is new. '*The sum of the squares on two sides of a triangle is double the sum of the squares on half the base and on the line joining the vertex to the middle point of the base.*' This, Mr. Wilson tells us, is 'Apollonius' Theorem ': but, even with that mighty name to recommend it, I cannot help thinking it rather more curious than useful.

Th. 13 (p. 96) is Euc., II. 9, 10.   Proved algebraically, and thus degraded from the position of a (fairly useful) geometrical Theorem to a mere addition-sum, of no more value than millions of others like it.

In the next proposition we suddenly transfer our allegiance, for no obvious reason, from Arabic to Latin numerals.

Problem 1 (p. 99) is Euc. I. 42 : old proof.

Pr. ii. (p. 100) is Euc. I. 44 : proved nearly as in Euclid, but labours under the same defect as Pr. 8 (p. 66) in that it assumes, without proof, Euc. Ax. 12.

Pr. iii (p. 100) is Euc. I. 45 : old proof.

Pr. iv (p. 101) is Euc. II. 14 : old proof.

Pr. v (p. 103) is new. '*To construct a rectilineal figure equal to a given rectilineal figure and having the number*

*of its sides one less than that of the given figure ; and thence
to construct a triangle equal to a given rectilineal figure.'*
This I have already noticed (see p. 167). It really is not
worth interpolating as a new Proposition. And its con-
cluding clause is, if I may venture on so harsh an expression, childish : it reminds me of nothing so much as the
Irish patent process for making cheap shoes—by taking
boots and cutting off the tops!

Pr. VI (p. 103) is '*To divide a straight line, either inter-
nally or externally, into two segments such that the rectangle
contained by the given line and one of the segments may be
equal to the square on the other segment.'* The case of
internal section is Euc. II. 11, with the old proof. The
other case is new, and worth interpolating.

I have now discussed, with as much care and patience
as the lateness of the hour will permit, so much of this
new Manual as corresponds to Euc. I, II, and I hope
your friends are satisfied.

[*A gentle cooing, as of satisfied ghosts, is heard in the air.*]

I will now give you in a few words the net result of
it all, and will show you how miserably small is the basis
on which Mr. Wilson and his coadjutors of the 'Association' rest their claim to supersede the Manual of
Euclid.

[*An angry moaning, as of ghosts suffering from neuralgia,
surges round the room, till it dies away in the chimney.*]

By breaking up certain of the Propositions of Euc. I, II, and including some of the Corollaries, we get 73 Propositions in all—57 Theorems and 16 Problems. Of these 73, this Manual omits 14 (10 Theorems and 4 Problems); it proves 43 (32 Theorems and 11 Problems) by methods almost identical with Euclid's; for 10 of them (9 Theorems and a Problem) it offers new proofs, against which I have recorded my protest, one being illogical, 2 (needlessly) employing 'superposition,' 2 deserting Geometry for Algebra, and the remaining 4 omitting the diagonals in Euc. II ; and finally it offers 6 new proofs, which I think may fairly be introduced as alternatives for those of Euclid.

In all this, and in all the matters previously discussed, I fail to see one atom of reason for abandoning Euclid. Have you any yet-unconsidered objections to urge against my proposal ' that the sequence and numeration of Euclid be kept unaltered ' ?

[*Dead silence is the only reply.*]

Carried, *nemine contragemente!* And now, Prisoner at the Bar (I beg your pardon, I should say ' Professor on the Sofa '), have you, and your attendant phantoms, any other reasons to urge for regarding this Manual as in any sense a substitute for Euclid's—as in any sense anything else than a revised edition of Euclid ?

Nie. We have nothing more to say.

Min. Then I can but repeat with regard to this new-born ' follower ' of the Syllabus, what I said of the

Syllabus itself. Restore the Problems (which are also Theorems) to their proper places; keep to Euclid's numbering (interpolating your new Propositions where you please); and your new book may yet prove a valuable addition to the literature of Elementary Geometry.

[*A tremulous movement is seen amid the ghostly throng. They waver fitfully to and fro, and finally drift off in the direction of one corner of the ceiling. When the procession has got well under weigh,* NIEMAND *himself becomes hazy, and floats off to join them. The whole procession gradually melts away into vacancy,* DIAMOND *going last, nibbling at the heels of* NERO, *for which a pair of gorgeous Roman sandals seem to afford but scanty protection.*]

ACT IV.

[*Scene as before. Time, the early dawn.* MINOS *slumbering uneasily, having fallen forwards upon the table, his forehead resting on the inkstand. To him enter* EUCLID *on tip-toe, followed by the phantasms of* ARCHIMEDES, PYTHAGORAS, ARISTOTLE, PLATO, &c., *who have come to see fair play.*]

§ 1. *Treatment of Pairs of Lines.*

Euc. Are all gone?

Min.   'Be cheerful, sir:
Our revels now are ended: these our actors,
As I foretold you, were all spirits, and
Are melted into air, into thin air!'

Euc. Good. Let us to business. And first, have you found any method of treating Parallels to supersede mine?

Min. No! A thousand times, no! The infinitesimal method, so gracefully employed by M. Legendre, is unsuited to beginners; the method by transversals has not yet been offered in a logical form: the 'equidistant' method is too cumbrous: and as for the method of 'direction,' it is simply a rope of sand—it breaks to pieces wherever you touch it!

Euc. We may take it as a settled thing, then, that you have found no sufficient cause for abandoning either my sequence of propositions or their numbering, and that all that now remains to be considered is whether any important modifications of my Manual are desirable?

Min. Most certainly.

Euc. Have you met with any striking novelty on the subject of a practical test for the meeting of Lines?

Min. There is *one* rival to your 12th Axiom which is formidable on account of the number of its advocates— the one usually called ' Playfair's Axiom.'

Euc. We have discussed that matter already (p. 42).

Min. But what have you to say to those who reject Playfair's axiom as well as yours?

Euc. I simply ask them what practical test, as to the meeting of two given finite Lines, they propose to employ. Not only will they find it necessary to prove, in certain theorems, that two given finite Lines will meet if produced, but they will even find themselves sometimes obliged to prove it of two Lines, of which the only geometrical fact known is that they possess the very property which forms the subject of my axiom. I ask them, in short, this question:—' Given two Lines making, with a certain transversal, two interior angles together less than two right angles, how do you propose to prove, without my axiom, that they will meet if produced?'

Min. The advocates of the ' direction ' theory would of course reply, ' We can prove, from the given property,

that they have different directions: and then we bring
in the axiom that Lines having different directions will
meet if produced.'

Euc. All *that* you have satisfactorily disposed of in your
review of Mr. Wilson's Manual.

Min. The only other substitute, that I know of, belongs
to the 'equidistant' theory, which replaces your axiom by
three or four new axioms and six new theorems. *That*
substitute, also, I have seen reason to reject.

My general conclusion is that your method of treatment of all these subjects is the best that has yet been
suggested.

Euc. Any noticeable innovations in the treatment of
Right Lines and Angles?

Min. Those subjects I should be glad to talk over with
you.

Euc. With all my heart. And now how do you propose to conduct this our final interview?

Min. I should wish, in the first place, to lay before you
the general charges which have been brought against you:
then to discuss your treatment of Lines and Angles, as
contrasted with that of your 'Rivals'; and lastly the
omissions, alterations, and additions proposed by them.

Euc. Good. Let us begin.

Min. I will take the general charges under three headings:—Construction, Demonstration, and Style. And first
as to Construction :—

§ 2. *Euclid's Constructions.*

I am told that you indulge too much in 'arbitrary restrictions.' Mr. Reynolds says (Pref. p. vi.) 'The arbitrary restrictions of Euclid involve him in various inconsistencies, and exclude his constructions from use. When, for instance, in order to mark off a length upon a straight line, he requires us to describe five circles, an equilateral triangle, one straight line of limited, and two of unlimited length, he condemns his system to a divorce from practice at once and from sound reason.'

Euc. Mr. Reynolds has misunderstood me: I do not require all that construction in Prop. 3. To explain my meaning I must go back to Prop. 2, and I must ask your patience while I make a few general remarks on construction. The machinery I allow consists of a pencil, a ruler, and a pair of compasses to be used for drawing a circle about a given centre and *passing through a given point* (that is what I mean by 'at any distance'), but *not* to be used for transferring distances from one part of a diagram to another *until it has been shown that such transference can be effected by the machinery already allowed.*

Min. But why not allow such transference without proving that possibility?

Euc. Because it would be introducing as a *postulate* what is really a *problem.* And I go on the general principle of never putting a problem among my postulates, nor a theorem among my axioms.

Min. I heartily agree in your general principle, though I need scarcely remind you that it has been frequently charged against you, as a *fault,* that you state as an axiom what is really a theorem.

Euc. That charge has been met (see p. 42). To return to my subject. I merely prove, once for all, in Prop. 2, that a line *can*‧be drawn, from a given point, and equal to a given line, by the original machinery alone, and *without* transferring distances. After that, my reader is welcome to transfer a distance by any method that comes handy, such as a bit of string &c.: and of course he may now transfer his compasses to a new centre. And this is all I expect him to do in Prop. 3.

Min. Then you *don't* expect these five circles &c. to be drawn whenever we have to cut off, from one Line, a part equal to another ?

Euc. Pas si bête, mon ami.

Min. Some of your Modern Rivals are, however, a little discontented with the very scanty machinery you allow.

Euc. 'A bad workman always‧quarrels with his tools.'

Min. Their charge against you is 'the exclusion of hypothetical constructions.' Mr. Wilson says (Pref. p. i.) 'The exclusion of hypothetical constructions may be mentioned as a self-imposed restriction which has made the confused order of his first book necessary, without any compensating advantage.'

Euc. In reply, I cannot do better than refer you to Mr. Todhunter's Essay on Elementary Geometry (p. 186). 'Confused order is rather a contradictory expression,' &c. (see p. 216).

Min. Your reply is satisfactory. Mr. Wilson himself is
an instance of the danger of such a method. Three times
at least (pp. 46, 70, 88) he produces Lines to meet without
attempting to prove that they *will* meet.

§ 3. *Euclid's Demonstrations.*

Min. The next heading is 'Demonstration.' You are
charged with an 'invariably syllogistic form of reasoning.'
(Wilson, Pref. p. i.)

Euc. Do you know, I am vain enough to think that
a merit rather than a defect? Let me quote what Mr.
Cuthbertson says on this point (Pref. p. vii.). 'Euclid's
mode of demonstration, in which the conclusion of each
step is preceded by reasoning expressed with all the exact-
ness of the minor premiss of a syllogism, of which some
previous proposition is the major premiss, has been adopted
as offering a good logical training, and also as being
peculiarly adapted for teaching large classes, rendering
it possible for the teacher to call first upon one, then
upon another, and so on, to take up any link in the
chain of argument.' Perhaps even Mr. Wilson's own
book would not be the worse if the reasoning were a
trifle more 'syllogistic'!

Min. A fair retort. You are also charged with 'too
great length of demonstration.' Mr. Wilson says (Pref.
p. i.) 'The real objections to Euclid as a text-book are . . .
the length of his demonstrations.' And Mr. Cooley says
(Pref. p. i.) 'The important and fertile theorems, which
crown the heights in this field of knowledge, are here

all retained, and those only are omitted which seem to be but the steps of a needlessly protracted ascent. The short road thus opened will be found perfectly solid in construction, and at the same time far less tedious and fatiguing than the circuitous one hitherto in vogue.'

Euc. I think Mr. Wilson's Th. 17 (p. 27), with its five figures (all necessary, though he only draws one), and still more his marvellous problem, ' approached by four stages,' which fills pages 69 to 72, are pretty good instances of lengthy demonstration. And Mr. Cooley's ' short and solid road' contains, if I remember right, a rather break-neck crevasse !

Min. The next charge against you is ' too great *brevity* of demonstration.' Mr. Leslie (a writer whom I have not thought it necessary to review as a ' Modern Rival,' as his book is nearly seventy years old) says (Pref. p. vi.) ' In adapting it ' (the Elements of Euclid) ' to the actual state of the science, I have . . . sought to enlarge the basis . . . The numerous additions which are incorporated in the text, so far from retarding will rather facilitate progress, by rendering more continuous the chain of demonstration. To multiply the steps of ascent, is in general the most expeditious mode of gaining a summit.'

Euc. I think you had better refer him to Mr. Wilson and Mr. Cooley : they will answer *him*, and he in his turn will confute *them* !

Min. The last charge relating to demonstration is, in Mr. Wilson's words (Pref. p. viii.) ' the constant reference to general axioms and general propositions, which are no clearer in the general statement than they are in the

particular instance,' which practice, he says, makes the
study of Geometry 'unnecessarily stiff, obscure, tedious
and barren.'

Euc. *One* advantage of making a general statement, and
afterwards referring to it instead of repeating it, is that
you have to go through the mental process of affirming or
proving the truth *once for all* : apparently Mr. Wilson
would have you begin *de novo* and think out the truth
every time you need it!   But the great reason for always
referring back to your universal, instead of affirming the
particular (Mr. Wilson is merely starting the old logical
hare ' Is the syllogism a *petitio principii*?'), is that the
truth of the particular does not rest on any data peculiar
to itself, but on general principles applicable to all similar
cases; and that, *unless those general principles prove the
conclusion for* all *cases, they cannot be warranted to prove
it for any one selected case.*   If, for instance, I see a
hundred men, and am told that some assertion is true
of ninety-nine of them, but am *not* told that it is true
of *all,* I am not justified in affirming it of any selected
man ; for he *might* chance to be the excepted one.   Now
the assertion, that the truth of the particular case under
notice depends on general principles, and not on peculiar
circumstances, is neither more nor less than the assertion
of the *universal* affirmative which Mr. Wilson deprecates.

§ 4. *Euclid's Style.*

Min. Quite satisfactory.   I will now take the third
heading, namely ' Style.'

You are charged with Artificiality, Unsuggestiveness, and Want of Simplicity. Mr. Wilson says (Pref. p. i.) 'The real objections to Euclid as a text-book are his artificiality . . . and his unsuggestiveness,' and again, 'he has sacrificed, to a great extent, simplicity and naturalness in his demonstrations, without any corresponding gain in grasp or cogency.'

Euc. Well, really I cannot deal with general charges like these. I prefer to abide by the verdict of my readers during these two thousand years. As to 'unsuggestiveness,' that is a charge which cannot, I admit, be retorted on Mr. Wilson : his book is *very* suggestive—of remarks which, perhaps, would not be *wholly* 'music to his ear' !

§ 5. *Euclid's treatment of Lines and Angles.*

Min. Let us now take the subjects of Right Lines and Angles; and first, the 'Right Line.'

I see, by reference to the original, that you define it as a line 'which lies evenly as to points on it.' That of course is only an attempt to give the mind a grasp of the idea. It leads to no geometrical results, I think ?

Euc. No : nor does *any* definition of it, that I have yet seen.

Min. I have no rival definitions to propose. Mr. Wilson's 'which has the same direction at all parts of its length' has perished in the collapse of the 'direction' theory : and M. Legendre's 'the shortest course from one point to another' is not adapted for the use of a beginner.

And I do not know that any change has been suggested in your test of a right Line in Prop. 14.

The next subject is ' Angles.'

Your definition would perhaps be improved, if for ' inclination to' we were to read ' declination from,' for, the greater the angle the greater the *de*clination, and the less (as it seems to me) the *in*clination.

Euc. I agree with you.

Min. The next point is that you limit the size of an angle to something less than the sum of two right angles.

Euc. What advantage is claimed for the extension of the definition ?

Min. It is a prospective rather than an immediate one. It must be granted you that the larger angles are not needed in the first four Books—

Euc. In the first *six* Books.

Min. Nay, surely you need them in the Sixth Book?

Euc. Where ?

Min. In Prop. 33, where you treat of ' any equimultiples whatever' of an angle, of an arc, and of a sector. You cannot possibly assume the multiple angle to be always less than two right angles.

Euc. You think, then, that a multiple of an angle must itself be an angle ?

Min. Surely.

Euc. Then a multiple of a man must itself be a man. If I contemplate a man as multiplied by the number ten thousand, I must realise the idea of a man ten thousand times the size of the first ?

Min. No, you need not do *that*.

Euc. Thanks : it *is* rather a strain on the imaginative faculty.

Min. You mean, then, that the multiple of an angle may be conceived of as so many separate angles, not in contact, nor added together into one ?

Euc. Certainly.

Min. But you have to contemplate the case where two such angular magnitudes are equal, and to infer from that, by III. 26, that the subtending arcs are equal. How can you infer this when your angular magnitude is not one angle but many ?

Euc. Why, the sum total of the first set of angles is equal to the sum total of the second set. Hence the second set can clearly be broken up and put together again in such amounts as to make a set equal, each to each, to the first set : and then the sum total of the arcs, and likewise of the sectors, will evidently be equal also.

But if you contemplate the multiples of the angles as single angular magnitudes, I do not see how you prove the equality of the subtending arcs : for *my* proof applies only to cases where the angle is less than the sum of two right angles.

Min. That is very true, and you have quite convinced me that we ought to observe that limit, and not contemplate ' angles of rotation ' till we enter on the subject of Trigonometry.

As to right angles, it has been suggested that your axiom ' all right angles are equal to one another ' is capable of proof as a theorem.

Euc. I do not object to the interpolation of such a

o

theorem, though there is very little to distinguish so simple a theorem from an axiom.

Min. Let us now consider the omissions, alterations, and additions, which have been proposed by your Modern Rivals.

§ 6. *Omissions, alterations, and additions,* *suggested by Modern Rivals.*

Euc. Which of my Theorems have my Modern Rivals proposed to omit ?

Min. Without dwelling on such extreme cases as that of Mr. Pierce, who omits no less than 19 of the 35 Theorems in your First Book, I may say that the only two, as to which I have found anything like unanimity, are I. 7 and II. 8.

Euc. As to I. 7, I have several reasons to urge in favour of retaining it.

First, it is useful in proving I. 8, which, without it, is necessarily much lengthened, as it then has to include *three* cases : so that its omission effects little or no saving of space.

Secondly, the modern method of proving I. 8 independently leaves I. 7 still unproved.

Min. That reason has no weight unless you can prove I. 7 to be valuable for itself.

Euc. True, but I think I *can* prove it ; for, thirdly, it shows that, of all plane figures that can be made by hingeing rods together, the *three*-sided ones (and these only) are *rigid* (which is another way of stating the fact that there cannot be *two* such figures on the same base). This is analogous to the fact, in relation to solids contained by plane surfaces hinged together, that *any* such solid is rigid, there being no maximum number of sides.

And fourthly, there is a close analogy between I. 7, 8 and III. 23, 24. These analogies give to Geometry much of its beauty, and I think that they ought not to be lost sight of.

Min. You have made out a good case. Allow me to contribute a 'fifthly.' It is one of the very few propositions that have a direct bearing on practical science. I have often found pupils much interested in learning that the principle of the rigidity of triangles is of constant use in architecture, and even in so homely a matter as the making of a gate.

The other Theorem which I mentioned, II. 8, is now so constantly ignored in examinations that it is very often omitted, as a matter of course, by students. It is believed to be extremely difficult and entirely useless.

Euc. Its difficulty has, I think, been exaggerated. Have you tried to teach it?

Min. I *have* occasionally found pupils amiable enough to listen to what they felt sure would be of no service in examinations. My experience has been wholly among undergraduates, any one of whom, if of average ability, would, I think, master it in from five to ten minutes.

Euc. No very exorbitant demand on your pupil's time. As to its being 'entirely useless,' I grant you it is of no *immediate* service, but you will find it eminently useful when you come to treat the Parabola geometrically.

Min. That is true.

Euc. Let us now consider the new methods of proof suggested by my Rivals.

Min. Prop. 5 has been much attacked—I may say trampled on—by your Modern Rivals.

Euc. Good. So that is why you call it 'The Asses' Bridge'? Well, how many new methods do they suggest for crossing it?

Min. One is 'hypothetical construction,' M. Legendre bisecting the base, and **Mr.** Pierce the vertical angle, but without any proof that the thing can be done.

Euc. So long as we agree that beginners in Geometry shall be limited to the use of Lines and Circles, so long will it be unsafe to assume a point as found, or a Line as drawn, merely because we are sure it *exists*. For example, it is axiomatic, of course, that every angle has a bisector: but it is equally obvious that it has two trisectors: and if I may assume the one as drawn, why not the others also? However we have discussed this matter already (p. 19).

Min. A second method is 'superposition,' adopted by Mr. Wilson and Mr. Cuthbertson—a method which here involves the *reversing* of the triangle, before applying it to its former position.

Euc. That also we have discussed (p. 46). What is the method adopted in the new Manual founded on the Syllabus of the Association?

Min. The same as Mr. Pierce's. Mr. Reynolds has a curious method: he treats the sides as obliques 'equally remote from the perpendicular.'

Euc. Curious, indeed.

Min. But perhaps the most curious of all is Mr. Willock's method : *he* treats the sides as radii of a circle, and the base as a chord.

Euc. He had better have made them asymptotes of a hyperbola at once ! *C'est magnifique, mais ce n'est pas la—Géométrie.*

Min. Two of your Rivals prove Prop. 8 from Prop. 24.

Euc. 'Putting the cart before the horse,' in my humble opinion.

Min. For a brief proof of Prop. 13, let me commend to your notice Mr. Reynolds'—consisting of the seven words ' For they fill exactly the same space.'

Euc. Why so lengthy ? The word 'exactly' is super-fluous.

Min. Instead of your chain of Theorems, 18, 19, 20, several writers suggest 20, 19, 18, making 20 axiomatic.

Euc. That has been discussed already (p. 55).

Min. Mr. Cuthbertson's proof of Prop. 24 is, if I may venture to say so, more complete than yours. He con-structs his figure without considering the relative lengths of the sides, and then proves the 3 possible cases separately.

Euc. I think it an improvement.

Min. There are no other noticeable innovations, that have not been already discussed, except that Mr. Cuthbertson proves a good deal of Book II by a quasi-algebraical method, without exhibiting to the eye the actual squares and rectangle : while Mr. Reynolds does it by pure algebra.

Euc. I think the actual squares, &c. most useful for be-ginners, making the theorems more easy to understand and to remember. *Algebraical* proofs of course introduce the difficulty of ' incommensurables.'

Min. We will now take the new propositions, &c. which have been suggested.

Here is an Axiom :—' *Two lines cannot have a common segment.*'

Euc. Good. I have tacitly assumed it, but it may as well be stated.

Min. Several new Theorems have been suggested, but only two of them seem to me worth mentioning. They are :—

' *All right angles are equal.*'

Euc. I have already approved of that (p. 193).

Min. The other is one that is popular with most of your Rivals :—

' *Of all the Lines which can be drawn to a Line from a point without it, the perpendicular is least; and, of the rest, that which is nearer to the perpendicular is less than one more remote; and the lesser is nearer than the greater; and from the same point only two equal Lines can be drawn to the other Line, one on each side of the perpendicular.*'

Euc. I like it on the whole, though so long an enunciation will be alarming to beginners. But it is strictly analogous to III. 7. Introduce it by all means in the revised edition of my Manual. It will be well, however, to lay it down as a general rule, that no proposition shall be so interpolated, unless it be of such importance and value as to be thought worthy of being quoted as proved, in the same way in which candidates in examinations are now allowed to quote propositions of mine.

Min. (*with a fearful yawn*) Well! I have no more to say.

§ 7. *The summing-up.*

Euc. 'The cock doth craw, the day doth daw,' and all
respectable ghosts ought to be going home. Let me carry
with me the hope that I have convinced you of the importance, if not the necessity, of retaining my order and numbering, and my method of treating straight Lines, angles,
right angles, and (most especially) Parallels. Leave me
these untouched, and I shall look on with great content-
ment while other changes are made—while my proofs are
abridged and improved—while alternative proofs are appended to mine—and while new problems and theorems are
interpolated.

In all these matters my Manual is capable of almost
unlimited improvement.

[*To the sound of slow music,* EUCLID *and the other ghosts*
'*heavily vanish,*' *according to Shakespeare's approved stage-
direction.* MINOS *wakes with a start, and betakes himself
to bed,* '*a sadder and a wiser man.*']

APPENDICES.

APPENDIX I.

Extract from Mr. Todhunter's essay on 'Elementary Geometry,' included in ' The Conflict of Studies, &c.'

IT has been said by a distinguished philosopher that England is " usually the last to enter into the general movement of the European mind." The author of the remark probably meant to assert that a man or a system may have become famous on the continent, while we are almost ignorant of the name of the man and the claims of his system. Perhaps, however, a wider range might be given to the assertion. An exploded theory or a disadvantageous practice, like a rebel or a patriot in distress, seeks refuge on our shores to spend its last days in comfort if not in splendour. Just when those who originally set up an idol begin to suspect that they have been too extravagant in their devotions we receive the discredited image and commence our adorations. It is a less usual but more dangerous illustration of the principle, if just as foreigners are learning to admire one of our peculiarities we should grow weary of it.

In teaching elementary geometry in England we have for a long time been accustomed to use the well-known *Elements of Euclid.* At the present moment, when we learn from the best testimony, namely, the admission of anti-Euclideans, that both in France and Italy dissatisfaction is felt with the system

hitherto used, accompanied with more or less desire to adopt
ours, we are urged by many persons to exchange our system
for one which is falling out of favour on the continent.

. . . .

Many assertions have been made in discussion which rest
entirely on the authority of the individual advocate, and thus it
is necessary to be somewhat critical in our estimate of the value
of the testimony.   Two witnesses who are put prominently
forward are MM. Demogeot and Montucci, who drew up a
report on English education for the French Government.   Now
I have no doubt that these gentlemen were suited in some
respects to report on English education, as they were selected
for that purpose; but I have searched in vain for any evidence
of their special mathematical qualifications.   No list of mathematical publications that I have consulted has ever presented
either of these names, and I am totally at a loss to conceive on
what grounds an extravagant respect has been claimed for their
opinions.   The following sentence has been quoted with appro-
bation from these writers : "Le trait distinctif de l'enseignement
des mathématiques en Angleterre c'est qu'on y fait appel plutôt
à la mémoire qu'à l'intelligence de l'élève."   In the first place
we ought to know on what evidence this wide generalisation is
constructed.   Did the writers visit some of the humbler schools
in England in which the elements of arithmetic and mensuration
were rudely taught, and draw from this narrow experience an
inference as to the range of mathematical instruction throughout
England ?   Or did they find on inspecting some of our larger
public schools that the mathematical condition was unsatisfactory?
In the latter case this might have arisen from exclusive devotion
to classics, or from preference for some of the fashionable novelties
of the day, or from want of attention and patience in the teachers.
On the most unfavourable supposition the condemnation pronounced on the general mathematical training in England cannot
be justified.   But take some kind of experimental test.   Let
an inquirer carefully collect the mathematical examination papers

issued throughout England in a single year, including those proposed at the Universities and the Colleges, and those set at the Military Examinations, the Civil Service Examinations, and the so-called Local Examinations. I say then, without fear of contradiction, that the original problems and examples contained in these papers will for interest, variety, and ingenuity surpass any similar set that could be found in any country of the world. Then any person practically conversant with teaching and examining can judge whether the teaching is likely to be the worst where the examining is the most excellent.

The sentence quoted from MM. Demogeot and Montucci, in order to have any value, ought to have proceeded from writers more nearly on a level with the distinguished mathematical teachers in England. So far as any foundation can be assigned for this statement, it will probably apply not to mathematics especially but to all our studies, and amount to this, that our incessant examinations lead to an over culti- vation of the memory. Then as to the practical bearing of the remark on our present subject it is obvious that the charge, if true, is quite independent of the text-book used for instruction, and might remain equally valid if Euclid were exchanged for any modern author.

The French gentlemen further on contrast what they call Euclid's verbiage with the elegant conciseness of the French methods. It is surely more than an answer to these writers to oppose the high opinion of the merits of Euclid expressed by mathematicians of European fame like Duhamel and Hoüel. See the *First Report of the Association for the Improvement of Geometrical Teaching*, p. 10.

When we compare the lustre of the mathematical reputation of these latter names with the obscurity of the two former, it seems that there is a great want of accuracy in the statement made in a recent circular: 'The opinion of French mathe- maticians on this question, is plainly expressed in the Report of MM. Demogeot and Montucci. . . .'

I should have to quote very largely indeed if I wished to draw attention to every hazardous statement which has been advanced; I must therefore severely restrain myself. Consider the following: 'Unquestionably the best teachers depart largely from his words, and even from his methods. That is, they use the work of Euclid, but they would teach better without it. And this is especially true of the application to problems. Everybody recollects, even if he have not the daily experience, how unavailable for problems a boy's knowledge of Euclid generally is.' The value of such a statement depends entirely on the range of the experience from which it has been derived. Suppose for instance that the writer had been for many years an examiner in a large University in which against each candidate's name the school was recorded from which he came; suppose that the writer had also been much engaged in the numerous examinations connected with the military institutions; suppose that he had also been for a quarter of a century in residence at one of the largest colleges at Cambridge, and actively employed in the tuition; suppose also that it had been his duty to classify the new students for lecture purposes by examining them in Euclid and other parts of elementary mathematics; and finally suppose that he was in constant communication with the teachers in many of the large schools: then his opinion would have enjoyed an authority which in the absence of these circumstances cannot be claimed for it.

If I may venture to refer to my own experience, which I fear commenced when the writer whom I have just quoted was in his cradle, I may say that I have taught geometry both Euclidean and non-Euclidean, that my own early studies and prepossessions were towards the latter, but that my testimony would now be entirely in favour of the former.

.   .   .   .

I admit that to teach Euclid requires patience both from the tutor and the pupil; but I can affirm that I have known many teachers who have succeeded admirably, and have sent a large

number of pupils to the University well skilled in solving
deductions and examples; nor have I ever known a really
able and zealous teacher to fail. I am happy to supplement
my own testimony by an extract from the very interesting
lecture on Geometrical Teaching by Dr. Lees, of St. Andrews.
' Whatever may be the cause of failure in England, it is clear
as any demonstration can be that the failure cannot be ascribed to
Euclid. Because in Scotland we do employ Euclid as the text-
book for our students, and in Scotland we have the teaching of
Geometry attended with the most complete success ; and this
not only in the colleges, but in all the higher and more important
schools and academies of the country, and in many of the parish
schools even, where the attention of the teacher is necessarily so
much divided.' See also the remarkable *Narrative-Essay on a
Liberal Education,* by the Rev. S. Hawtrey, A.M., Assistant-
Master, Eton.

During the existence of the East India Company's military
college at Addiscombe, it is well known that the cadets were
instructed in mathematics by the aid of a course drawn up by
the late Professor Cape. The geometry in this course was of
the kind which our modern reformers recommend, being founded
on Legendre, and adopting the principle of hypothetical constructions which is now so emphatically praised. In certain
large schools where youths were trained for the military colleges
it was usual to instruct a class of candidates for Woolwich, in
Euclid, and a class of candidates for Addiscombe in Cape's
adaptation of Legendre. Fairness in the procedure was secured
by giving the same number of hours by the same masters to
each class ; and the honour and rewards which attended success
supplied an effectual stimulus both to teachers and pupils. Now
consider the result. I was assured by a teacher who was for
many years distinguished for the number and the success of his
pupils, that the training acquired by the Euclid class was far
superior to that acquired by the Legendre class. The Euclid

was not more difficult to teach and was more potent and more beneficial in its influence. The testimony made the stronger impression on me because at the time I was disposed from theoretical considerations to hold an opposite opinion; I was inclined for example to support the use of hypothetical constructions. Such experience as I afterwards gained shewed the soundness of the judgment at which the practical teacher had arrived; and I have also received the emphatic evidence of others who had good opportunities of considering the question, and had come to the same conclusion. I have myself examined at Woolwich and at Addiscombe, and am confident that the teaching in both institutions was sound and zealous; but I have no hesitation in saying that the foundation obtained from Euclid was sounder than that from Legendre.

. . . .

Although I have admitted that the study of Euclid is one that really demands patient attention from the beginner, yet I cannot admit that the tax is unreasonable. My own experience has been gained in the following manner. Some years since on being appointed principal mathematical lecturer in my college, more systematic arrangements were introduced for the lectures of the freshmen than had been previously adopted; and as the Euclid seemed to be one of the less popular subjects I undertook it myself. Thus for a long period the way in which this has been taught in schools, and the results of such teaching, have been brought under my notice. It need scarcely be said that while many of the students who have thus presented themselves to me have been distinguished for mathematical taste and power, yet the majority have been of other kinds; namely, either persons of ability whose attention was fully occupied with studies different from mathematics, or persons of scanty attainments and feeble power who could do little more than pass the ordinary examination. I can distinctly affirm that the cases of hopeless failure in Euclid were very few; and the advantages derived from the study, even by men of feeble ability, were most decided.

In comparing the performance in Euclid with that in Arithmetic and Algebra there could be no doubt that the Euclid had made the deepest and most beneficial impression : in fact it might be asserted that this constituted by far the most valuable part of the whole training to which such persons were subjected. Even the modes of expression in Euclid, which have been theoretically condemned as long and wearisome, seemed to be in practice well adapted to the position of beginners. As I have already stated there appears to me a decided improvement gradually taking place in the knowledge of the subject exhibited by youths on entering the University. My deliberate judgment is that our ordinary students would suffer very considerably if instead of the well-reasoned system of Euclid any of the more popular but less rigid manuals were allowed to be taken as a substitute.

Let me now make a few remarks on the demand which has been made to allow other books instead of Euclid in *examinations*. It has been said : " We demand that we should not be,— as we are now, by the fact of Euclid being set as a text-book for so many examinations,—practically obliged to adhere to one book. Surely such a request, made by men who know what they want, and are competent to form an opinion on the subject,— and made in earnest,—should induce the Universities and other examining bodies to yield their consent. The grounds of the demand then are three ; that it is made in earnest, that it is made by those who know what they want, and that it is made by those who are competent to form an opinion on the subject. I need not delay on two of the grounds ; the experience of every day shews that claimants may know what they want, and be terribly in earnest in their solicitations, and yet it may be the duty of those to whom the appeal is made to resist it. Moreover it is obvious that the adoption of Euclid as a text book is pre- scribed by those who are equally in earnest and know what they recommend. In short if no institution is to be defended when it is attacked knowingly and earnestly, it is plain that no institu- tion is safe.

I turn then to the other ground, namely that the demand is made by men who are competent to form an opinion on the subject. Now it is not for me to affect to speak in the name of the University of Cambridge; mine is the opinion of only a private unofficial resident. But I have little doubt that many persons here will maintain, without questioning the competence of the claimants to form an opinion, that we ourselves are still much more competent to form an opinion.

For it will not be denied that in all which relates to mathematical knowledge we have an aggregate of eminence which far surpasses what has yet been collected together to press the demand on the University. Moreover as inspectors and judges we occupy a central position as it were, and thus enjoy opportunities which do not fall to isolated teachers however eminent and experienced. The incessant demands made upon the University to furnish examiners for schools and for the local examinations keep us as a body practically familiar with the standard of excellence attained in various places of instruction. Then as college lecturers and private tutors we have the strongest motives for keenly discriminating the state of mathematical knowledge in different schools, as shewn by the performance of the candidates when brought under our notice. Moreover some of the residents in the University by continued intercourse with old pupils, now themselves occupying important positions as teachers, are enabled to prolong and enlarge the experience which they may have already obtained directly or indirectly. If it is obvious that certain teachers by ability and devotion have for many years sent up well-trained pupils, the University may well consider that it would be neither right nor wise to deprive its best friends of their justly earned distinction, by relaxing in any way the rigour of the examinations. Instead then of urging an instant acquiescence with demands on the ground that those who make them are well qualified to judge, the claimants should endeavour by *argument* to convince others who are still better qualified to judge.

Here let me invite attention to the following remark which
has been made in support of the claim: "In every other subject
this is freely accorded ; we are not obliged to use certain
grammars or dictionaries, or one fixed treatise on arithmetic,
algebra, trigonometry, chemistry, or any other branch of science.
Why are we to be tied to one book in geometry alone ?" Now
in the first place it may be said that there are great advantages
in the general use of one common book; and that when one
book has long been used almost exclusively it would be rash to
throw away certain good in order to grasp at phantasmal benefits.
So well is this principle established that we have seen in recent
times a vigorous, and it would seem successful effort, to secure
the use of a common Latin Grammar in the eminent public
schools. In the second place the analogy which is adduced in the
remark quoted above would be rejected by many persons as involving an obvious fallacy, namely that the word *geometry* denotes
the same thing by all who use it. By the admirers of Euclid
it means a system of demonstrated propositions valued more
for the process of reasoning involved than for the results obtained.
Whereas with some of the modern reformers the rigour of the
method is of small account compared with the facts themselves.
We have only to consult the modern books named in a certain
list, beginning with the *Essentials of Geometry,* to see that
practically the object of some of our reformers is not to teach
the same subject with the aid of a different text-book, but to
teach something very different from what is found in Euclid,
under the common name of geometry.

It may be said that I am *assuming* the point in question,
namely, that Euclid is the best book in geometry; but this is
not the case. I am not an advocate for *finality* in this matter;
though I do go so far as to say that a book should be *decidedly*
better than Euclid before we give up the advantages of uni-
formity which it will be almost impossible to secure if the
present system is abandoned. But, as it has been well observed
by one of the most distinguished mathematicians in Cambridge,

P

" The demand is unreasonable to throw aside Euclid in favour of any compendium however meagre and however unsound; and this is really the demand which is made: it will be time enough to consider about the discontinuance of Euclid when a better book is deliberately offered." It may be added that the superiority to Euclid must be established by indisputable evidence; not by the author's own estimation, the natural but partial testimony of parental fondness; not by the hasty prediction of some anonymous and irresponsible reviewer; not by the authority of eminent men, unless the eminence is founded on mathematical attainments; not even by the verdict of teachers who are not conspicuous for the success of their pupils. The decision must rest with students, teachers, and examiners, of considerable reputation in the range of the mathematical sciences.

It must be allowed that there is diversity of opinion among the opponents of Euclid, for while the majority seem to claim freedom in the use of any text-books they please, others rather advocate the construction and general adoption of a new text-book. The former class on the whole seem to want something easier and more popular than Euclid; among the latter class there are a few whose notion seems to be that the text-book should be more rigorous and more extensive than Euclid. There are various considerations which seem to me to indicate that if a change be made it will not be in the direction of *greater rigour;* the origin of the movement, the character of the text-books which have hitherto been issued, and the pressure of more modern and more attractive studies, combine to warn us that if the traditional authority which belongs to Euclid be abandoned, geometry will be compelled to occupy a position in general education much inferior to that which it now holds.

. . . .

There is one very obvious mode of advancing the cause of the anti-Euclidean party, which I believe will do far more for them than the most confident assertions and predictions of the merits of the course which they advocate : let them train youths on their system to gain the highest places in the Cambridge Mathematical Tripos, and then other teachers will readily follow in the path thus opened to distinction. But it may naturally be said that as long as Euclid is prescribed for the text-book, the conditions of competition are unfair towards those who adopt some modern substitute; I will examine this point. In the Cambridge Examination for Mathematical Honours there are at present sixteen papers; a quarter of the first paper is devoted to book-work questions on Euclid. Now suppose that 1000 marks are assigned to the whole examination, and that about five of these fall to the book-work in Euclid. A student of any modern system would surely be able to secure some of these five marks, even from a stern Euclidean partisan. But to take the worst case, suppose the candidate deliberately rejects all chance of these five marks, and turns to the other matter on the paper, especially to the problems; here the advantage will be irresistibly on his side owing to the " superiority of the modern to the ancient methods of geometry " which is confidently asserted. It must be remembered that in spite of all warning and commands to the contrary, examiners will persist in making their papers longer than can be treated fully in the assigned time, so that the sacrifice of the book-work will be in itself trifling and will be abundantly compensated by the greater facility at the solution of problems which is claimed for the modern teaching, as compared with the " unsuggestiveness " of Euclid, and by the greater accuracy of reasoning, since we are told that " the logical training to be got from Euclid is very imperfect and in some respects bad." Thus on the whole the disciple of the modern school will even in the first paper of the Cambridge Tripos Examination be more favourably situated than the student of Euclid; and of course in the other papers the advantages in his favour become largely increased. For we must remember that we are expressly told that Euclid is " an unsuitable preparation for the higher mathematical training of the present day ; " and that " those who continue their mathematical reading with a view of obtaining

honours at the University . . . will gain much through economy
of time and the advantage of modern lights."

The final result is this; according to the promises of the
geometrical reformers, one of their pupils might sacrifice five
marks out of a thousand, while for all the remaining 995 his
chance would be superior to that of a Euclid-trained student. It
may be added that in future the Cambridge Mathematical
Examinations are to be rather longer than they have been up
to the date of my writing; so that the advantage of the anti-
Euclidean school will be increased. Moreover we must remember
that in the Smith's Prizes Examination the elementary geometry
of Euclid scarcely appears, so that the modern reformers would
not have here any obstacle to the triumphant vindication of
their superiority as teachers of the higher mathematics. The
marvellous thing is that in these days of competition for educa-
cational prizes those who believe themselves to possess such a
vast superiority of methods do not keep the secret to themselves, instead of offering it to all, and pressing it on the reluc-
tant and incredulous. Surely instead of mere *assertion* of the
benefits to be secured by the modern treatment, it will be far
more dignified and far more conclusive to *demonstrate* the proposition by brilliant success in the Cambridge Mathematical
Tripos. Suppose we were to read in the ordinary channels of
information some such notice as this next January : " The first
six wranglers are considered to owe much of their success to the
fact that in their training the fossil geometry of Alexandria was
thrown aside and recent specimens substituted ; " then opposition
would be vanquished, and teachers would wonder, praise, and
imitate. But until the promises of success are followed by a
performance as yet never witnessed we are reminded of the case
of a bald hairdresser who presses on his customers his infallible
specific for producing redundant locks.

. . . .

To those who object to Euclid as an inadequate course of
plane geometry it may then be replied briefly that it is easy,

if thought convenient or necessary, to supply any additional matter. But for my part I think there are grave objections to any large increase in the extent of the course of synthetical geometry which is to be prepared for examination purposes. One great drawback to our present system of mathematical instruction and examination is the monotony which prevails in many parts. When a mathematical subject has been studied so far as to master the essential principles, little more is gained by pursuing these principles into almost endless applications. On this account we may be disposed to regard with slender satisfaction the expenditure of much time on geometrical conic sections ; the student seems to gain only new facts, but no fresh ideas or principles. Thus after a moderate course of synthetical geometry such as Euclid supplies, it may be most advantageous for the student to pass on to other subjects like analytical geometry and trigonometry which present him with ideas of another kind, and not mere repetitions of those with which he is already familiar.

.   .   .   .

It has been said, and apparently with great justice, that examination in elementry geometry under a system of unrestricted text-books will be a very troublesome process ; for it is obvious that in different systems the demonstration of a particular proposition may be more or less laborious, and so may be entitled to more or fewer marks. This perplexity is certainly felt by examiners as regards geometrical conic sections ; and by teachers also who may be uncertain as to the particular system which the examiners may prefer or favour. It has been *asserted* that the objection thus raised is imaginary, and that " the manuals of geometry will not differ from one another nearly so widely as the manuals of algebra or chemistry : yet it is not difficult to examine in algebra and chemistry." But I am unable to feel the confidence thus expressed. It seems to me that much more variety may be expected in treatises on geometry than on algebra ; certainly if we may judge from the experience of the examiners at Cambridge the subject of geometrical conics is the most embarrassing which occurs at present, and this fact suggests a conclusion very different from that which is laid down in the preceding quotation. Of course there will be no trouble in examining a single school, because the system there adopted will be known and followed by the examiner.

I have no wish to exaggerate the difficulty; but I consider it to be real and serious, more especially as it presents itself at the outset of a youth's career, and so may cause disappointment just when discriminating encouragement is most valuable. But I think the matter must be left almost entirely to the discretion of examiners; the attempts which have been made to settle it by regulation do not seem to me very happy. For example, I read: "As the existing text-books are not very numerous, it would not be too much to require examiners to be acquainted with them sufficiently for the purpose of testing the accuracy of written, or even, if necessary, of oral answers." The language seems to me truly extraordinary. Surely examiners are in general men of more mathematical attainments than this implies; for it would appear that all we can expect them to do is to turn to some text-book and see if the student has correctly reproduced it. The process in a *viva voce* examination would be rather ignominious if when an answer had been returned by a candidate some indifferent manual had to be consulted to see if the answer was correct.

I have heard that an examining board has recently issued instructions to its executive officers to make themselves acquainted with the various text-books. This does not enjoin distinctly, what the above quotation implies, that the examiner is to accept all demonstrations which are in print as of nearly equal value; but it seems rather to suggest such a course. The point is important and should be settled. Suppose a candidate offered something taken from the *Essentials of Geometry*, and the examiner was convinced that the treatment was inadequate or unsound; then is the candidate nevertheless to obtain full marks? Again, it may be asked, why printed books alone are to be accepted; and why a student who has gone through a manuscript course of geometry should be precluded from following it? The regulation might be made that he should submit a copy of his manuscript course to the examiner in order that it might be ascertained whether he had reproduced it accurately. As I have already intimated, the only plan which can be adopted is to choose able and impartial men for examiners, and trust them to appreciate the merits of the papers submitted by the candidate to them.

The examiners will find many perplexing cases I have no doubt; one great source of trouble seems to me to consist in the fact that what may be a sound demonstration to one person with adequate preliminary study is not a demonstration to another person who has not gone through the discipline. To take a very simple example : let the proposition be, *The angles at the base of an isosceles triangle are equal.* Suppose a candidate dismisses this briefly with the words, *this is evident from symmetry*; the question will be, what amount of credit is to be assigned to him. It is quite possible that a well-trained mathematician may hold himself convinced of the truth of the proposition by the consideration of symmetry, but it does not follow that the statement would really be a demonstration for an early student. Or suppose that an another imbued with "the doctrine of the imaginary and inconceivable" says as briefly " the proposition is true, for the inequality of the angles is inconceivable and therefore false ; " then is the examiner to award full marks, even if he himself belongs to the school of metaphysics which denies that the inconceivable is necessarily the false ?

.   .   .   .

It has been urged as an objection against Euclid that the number of his propositions is too great. Thus it has been said that the 173 propositions of the six books might be reduced to 120, and taught in very little more than half the time required

to go through the same matter in Euclid. So far as the *half
time* is concerned this seems to be only an expression of belief
as to the result of an untried experiment; it is based on the
comparison of a few other books with Euclid, one of these being
the Course of the late Professor Cape ; as I have already stated,
actual experience suggests a conclusion directly contrary to
the present prediction. As to the *number* of propositions we
readily admit that a reduction might be made, for it is obvious
that we may in many cases either combine or separate according
to our taste. But the difficulty of a subject does not vary
directly as the number of propositions in which it is contained;
a single proposition will in some cases require more time and
attention than half a dozen others. I have no doubt that the
mixture of easy propositions with the more difficult is a great
encouragement to beginners in Euclid ; and instead of diminish-
ing the number of propositions I should prefer to see some
increase : for example I should like to have Euclid i. 26 divided
into two parts, and Euclid i. 28 into two parts.

Again, it has been said that Euclid is artificial, and that he
" has sacrificed to a great extent simplicity and naturalness in
his demonstrations ;" it is a curious instance of the difference of
opinion which we may find on the same subject, for, with a much
wider experience than the writer whom I quote, I believe that
Euclid maintains, and does not sacrifice, simplicity and naturalness in his system, assuming that we wish to have strictness
above all things.

The exclusion of hypothetical constructions has been represented as a great defect in Euclid ; and it has been said that this
has made the *confused order* of his first book necessary. Confused order is rather a contradictory expression; but it may be
presumed that the charge is intended to be one of confusion : I
venture to deny the confusion. I admit that Euclid wished to
make the subject depend on as few axioms and postulates as
possible ; and this I regard as one of his great merits ; it has
been shown by one of the most distinguished mathematicians of

our time how the history of science teaches in the clearest
language that the struggle against self-imposed restrictions
has been of the most signal service in the advancement of
knowledge.

The use of hypothetical constructions will not present itself
often enough to produce any very great saving in the demonstrations; while the difficulty which they produce to many
beginners, as shown by the experience to which I have already
referred, is a fatal objection to them. Why should a beginner
not assume that he can draw a circle through four given points
if he finds it convenient?

. . . .

Finally, I hold that Euclid, in his solution of the problems he
requires, supplies matter which is simple and attractive to
beginners, and which therefore adds practically nothing to their
labours, while it has the advantage of rendering his treatise far
more rigorous and convincing to them.

The objections against Euclid's order seem to me to spring
mainly from an intrusion of natural history into the region of
mathematics; I am not the first to print this remark, though it
occurred to me independently. It is to the influence of the
classificatory sciences that we probably owe this notion that it is
desirable or essential in our geometrical course to have all the
properties of triangles thrown together, then all the properties
of rectangles, then perhaps all the properties of circles; and so
on. Let me quote authority in favour of Euclid, far more
impressive than any which on this point has been brought
against him: " Euclid . . . fortunately for us, never dreamed of
a geometry of triangles as distinguished from a geometry of circles,
. . . but made one help out the other as he best could."

Euclid has been blamed for his adherence to the syllogistic
method; but it is not necessary to say much on this point,
because the reformers are not agreed concerning it: those who
are against the syllogism may pair off with those who are for
the syllogism. We are told in this connexion that, " the result

is, as every one knows, that boys may have worked at Euclid for
years, and may yet know next to nothing of Geometry." We
may readily admit that such may be the case with boys excep-
tionally stupid or indolent; but if any teacher records this as
the average result of his experience, it must I think be singularly
to his own discredit.

There is, I see, a notion that the syllogistic form, since it
makes the demonstrations a little longer, makes them more
difficult; this I cannot admit. The number of words employed
is not a test of the difficulty of a demonstration. An examiner,
especially if he is examining *viva voce*, can readily find out where
the difficulties of the demonstrations really lie; my own experience leads me to the conclusion that the syllogistic form instead
of being an impediment is really a great assistance, especially to
early students.

"Unsuggestiveness" has been urged as a fault in Euclid;
which is interpreted to mean that it does not produce ability to
solve problems. We are told: " Everybody recollects, even if he
have not the daily experience, how unavailable for problems a
boy's knowledge of Euclid generally is. Yet this is the true
test of geometrical knowledge; and problems and original work
ought to occupy a much larger share of a boy's time than they
do at present." I need not repeat what I have already said,
that English mathematicians, hitherto trained in Euclid, are
unrivalled for their ingenuity and fertility in the construction
and solution of problems. But I will remark that in the important mathematical examinations which are conducted at
Cambridge the rapid and correct solution of problems is of
paramount value, so that any teacher who can develop that
power in his pupils will need no other evidence of the merits
of his system.

Euclid's treatment of proportion has been especially marked
out for condemnation; indeed, with the boldness which attaches
to many assertions on the subject of elementary geometry, it
has been pronounced already *dead.* In my own college it has

never been laid aside ; only a few months since one of our most influential tutors stated that he was accustomed to give a proposition out of the fifth book of Euclid to some candidates for emoluments, and he considered it a very satisfactory constituent of the whole process of testing them.

I should exceedingly regret the omission of the fifth book of Euclid, which I hold to be one of the most important parts of the training supplied by Elementary Geometry. I do not consider it necessary for beginners to go through the entire book ; but the leading propositions might be mastered, and the student led to see how they can be developed if necessary. I may refer here to some valuable remarks which have been made on the subject by the writer of a *Syllabus of Elementary Geometry* . . . who himself I believe counts with the reformers. He sums up thus ". . . any easy and unsatisfactory short cuts (and I have sometimes seen an inclination for such) should be scouted, as a simple deception of inexperienced students."

However, I must remark that I see with great satisfaction the following *Resolution* which was adopted at a recent meeting of the *Association for the Improvement of Geometrical Teaching* : " That no treatise on Geometry can be regarded as complete without a rigorous treatment of ratio and proportion either by Euclid's method, or by some equally rigorous method of limits." It would be injudicious to lay much stress on resolutions carried by a majority of votes ; but at least we have a striking contradiction to the confident statement that Euclid's theory of proportion is *dead*. We shall very likely see here, what has been noticed before, that a course may be proposed which differs widely from Euclid's, and then, under the guidance of superior knowledge and experience, the wanderers are brought back to the old path. Legendre's return to Euclid's treatment of parallels is a conspicuous example ; see the valuable paper by Professor Kelland on *Superposition* in the *Edinburgh Transactions*, Vol. XXI.

I cannot avoid noticing one objection which has been urged

against Euclid in relation to his doctrine of proportion ; namely, that it leaves " the half-defined impression that all profound reasoning is something far fetched and artificial, and differing altogether from good clear common sense." It appears to me that if a person imagines that " good clear common sense" will be sufficient for mastering pure and mixed mathematics, to say nothing of contributing to their progress,—the sooner he is undeceived the better. Mathematical science consists of a rich collection of investigations accumulated by the incessant labour of many years, by which results are obtained far beyond the range of unassisted common sense; of these processes Euclid's theory of proportion is a good type, and it may well be said that from the degree of reverent attention which the student devotes to it, we may in most cases form a safe estimate of his future progress in these important subjects.

.   .   .   .

In conclusion I will say that no person can be a warmer advocate than I am for the *improvement of Geometrical Teaching* : but I think that this may be obtained without the hazardous experiment of rejecting methods, the efficacy of which a long experience has abundantly demonstrated.

APPENDIX II.

Extract from Mr. De Morgan's review of Mr. Wilson's Geometry, in the 'Athenæum' for July 18, 1868.

The Schools' Inquiry Commission has raised the question whether Euclid be, as many suppose, the best elementary treatise on geometry, or whether it be a mockery, delusion, snare, hindrance, pitfall, shoal, shallow, and snake in the grass.

.   .   .   .

We pass on to a slight examination of Mr. Wilson's book. We specially intend to separate the logician from the geometer. In the author's own interest, and that he may be as powerful a defender as can be of a cause we expect and desire to see fully argued, we recommend him to revise his notions of logic. We know that mathematicians care no more for logic than logicians for mathematics. The two eyes of exact science are mathematics and logic : the mathematical sect puts out the logical eye, the logical sect puts out the mathematical eye; each believing that it sees better with one eye than with two. The consequences are ludicrous.   On the one side we have, by confusion of words, the great logician Hamilton bringing forward two quantities which are 'one and the same quantity,' Breadth and Depth, while, within a few sentences, 'the greater the Breadth, the less the Depth.'   On the other side, we have the great mathematician, Mr. Wilson, also by confusion of

words, speaking of the 'invariably syllogistic form of his
[Euclid's] reasoning,' and, to show that this is not a mere
slip, he afterwards talks of the 'detailed syllogistic form' as
a 'source of obscurity to beginners, and damaging to true
geometrical freedom and power.'

Euclid a book of syllogistic form! We stared. We never
heard of such a book, except the edition of Herlinus and
Dasypodius (1566), who, quite ignorant that Euclid was
syllogistic already, made him so, and reckoned up the syllogisms.
Thus i. 47 has 'syllogismi novem' at the head. They did
not get much thanks; the book was never reprinted, and was
in oblivion-dust when Hamilton mentioned the zealous but
thick-headed logicians, as he called them. Prof. Mansel, in
our own day, has reprinted one of their propositions as a
curiosity. In 1831, Mr. De Morgan, advocating the reduction
of a few propositions to detailed syllogistic form as an exercise
for students, gave i. 47 as a specimen, in the Library of Useful
Knowledge; this was reprinted, we believe, in the preface to
various editions of Lardner's Euclid. A look will show the
difference between Euclid and syllogistic form. Had the
Elements been syllogistic, it would have been quoted in all
time, as a proof of the rapid diffusion of Aristotle's writings,
that they had saturated his junior contemporary with their
methods: with a controversy, most likely, raised by those who
would have contended that Euclid invented syllogistic form for
himself. Now it is well known that diffusion of Aristotle's
writings commenced after his death, and that it was—not quite
correctly—the common belief that evulgation did not take place
until two hundred years after his death. Could this belief
ever have existed if Euclid had invariably used 'syllogistic
form'?

What could have been meant? Craving pardon if wrong,
we suspect Mr. Wilson to mean that Euclid did not deal in
arguments with *suppressed premises.* Euclid was quite right:
the first reasonings presented to a beginner should be of full

statement. He may be trained to suppression : but the true way to abbreviation is from the full length. Mr. Wilson does not use the phrases of reasoning consistently. He tells the student that a *corollary* is 'a geometrical truth easily deducible from a theorem': and then, to the theorem that only one perpendicular can be drawn to a straight line, he gives as a corollary that the external angle of a triangle is greater than the internal opposite. This is not a corollary from the theorem, but a matter taken for granted in proving it.

Leaving this, with a recommendation to the author to strengthen his armour by the study of logic, we pass on to the system. There is in it one great point, which brings down all the rest if it fall, and may perhaps—but we must see Part II. before we decide—support the rest if it stand. That point is the treatment of the angle, which amounts to this, that certain notions about *direction*, taken as self-evident, are permitted to make all about angles, parallels and all, immediate consequences. The notion of continuous change, and consequences derived from it, enters without even an express assumption : ' continually ' is enough.

Mr. Wilson would not have ventured expressly to postulate that when a magnitude changes continuously, all magnitudes which change with it also change continuously. He knows that when a point moves on a line, an angle may undergo a sudden change of two right angles. He trusts to the beginner's perception of truth in the case before him : the whole truth would make that beginner feel that he is on a foundation of general principles made safe for him by selection, and only safe because the exceptions are not likely to occur to his mind. On this we write, as Newton wrote on another matter, *Falsa ! Falsa ! Non est Geometria !*

What ' direction ' is we are not told, except that ' straight lines which meet have different directions.' Is a direction a magnitude ? Is one direction greater than another ? We should suppose so ; for an angle, a magnitude, a thing which is to be halved and quartered, is the 'difference of the direction' of 'two straight lines that meet one another.' A better definition follows; the 'quantity of turning' by which we pass from one direction to another. But hardly any use is made of this, and none at the commencement. And why two definitions? Is the difference of two directions the *same thing* as the rotation by which we pass from one to the other? Is the difference of position of London and Rugby a number of miles on the railroad? Yes, in a loosely-derived and popular and slip-slop sense: and in like manner we say that one man *is* a pigeon-pie, and another *is* a shoulder of lamb, when we describe their contributions to a pic-nic. But *non est geometria !* Metaphor ·and paronomasia can draw the car of poetry; but they tumble the waggon of geometry into the ditch.

Parallels, of course, are lines which have the same direction. It is stated, as an immediate consequence, that two lines which meet cannot make the same angle with a third line, on the same side, for they are in different directions. Parallels are knocked over in a trice. There is a covert notion of direction, which, though only defined with reference to lines which meet, is straightway transferred to lines which do not. According to the definition, direction is a relation of lines which *do* meet, and yet lines which have the same direction can be lines which *never* meet. There is a great *quantity of turning* wanted; turning of implied assumption into expressed. Mr. Wilson would, we have no doubt, immediately introduce and defend all we ask for; and we quite admit that his system has a right to it. How do you know, we ask, that lines which have the same direction never meet? Answer—lines which meet have *different* directions. We know they have; but how do we know that, under the definition given, the relation called *direction* has any application at all to lines which never meet? The use of the notion of limits may give an answer : but what is the system of geometry which introduces continuity and limits to the mind as yet untaught to think of space and of

magnitude ?   Answer, a royal road.   If the difficulty were
met by expressed postulates, the very beginner himself would
be frightened.

There is a possibility that Mr. Wilson may mean that lines
which make the same angle with a third on the same side are
in the same direction. If this be the case, either he assumes
that lines equally inclined to one straight line are equally
inclined to all,—and this we believe he does, under a play on
the word 'direction'; or he makes a quibble only one degree
above a pun on his own arbitrary assumption of his right to
the word 'same': and this we do not believe he does.   He
should have been more explicit : he should have said, My system
involves an assumption which has lain at the root of many
attempts upon the question of parallels, and has always been
scouted as soon as seen. He should have added, I assume
Euclid's eleventh axiom : I have a notion of direction; I tell
you that lines which meet have different directions; I imply
that lines which make different angles with a third have also
different directions ; and I assume that lines of different directions
will meet.   Mr. Wilson is so concise that it is not easy to be
very positive as to how much he will admit of the above, or
how he will get over or round it. When put upon his
defence he must be more explicit. Mr. Wilson gives four
explicit axioms about the straight line : and not one about the
angle.

.   .   .   .

We feel confidence that no such system as Mr. Wilson has
put forward will replace Euclid in this country.   The old
geometry is a very English subject, and the heretics of this
orthodoxy are the extreme of heretics : even Bishop Colenso
has written a Euclid.   And the reason is of the same kind as
that by which the classics have held their ground in education.
There is a mixture of good sense and of what, for want of a
better name, people call prejudice : but to this mixture we owe
our stability.   The proper word is *postjudice*, a clinging to past

experience, often longer than is held judicious by after times. We only desire to avail ourselves of this feeling until the book is produced which is to supplant Euclid; we regret the manner in which it has allowed the retention of the faults of Euclid; and we trust the fight against it will rage until it ends in an amended form of Euclid.

APPENDIX III.

The enunciations of Tables I–IV, stated in full: with references to the various writers who have assumed or proved them.

[N.B.—Contranominals are bracketed. 'A' = 'assumed'; 'P' = 'proved.']

TABLE I. [See p. 32.]

	EUCLID	LEGENDRE	CUTHBERTSON	WILSON	PIERCE	WILLOCK	'SYLLABUS'-MANUAL	CHAUVENET	LOOMIS	MORELL	REYNOLDS	WRIGHT
1		P	P	A	P		A	A	P	A		
2 (1)	A	A	A			P		A	A	A		A
(2)						P						

1. $cc \,\mathbb{P}\, C.$ 'Lines, which have two common points, are ' coincidental.'

2 (1). $s \,\mathbb{P}\, (cc)'$ or S_{-1} ; $\equiv cc$ unique.   'Different Lines have not two common points; i.e. they have all points, but one, separate.' (Nothing is asserted of the excepted point.) This $\equiv$ 'Through two given points only one Line can be drawn.'

2 (2). $cs \,\mathbb{P}\, X.$ 'Lines, which have a common point and a separate point,' (or 'Different Lines, which have a common point,') ' are intersectional; i.e. have one point common and all others separate.'

TABLE I (*continued*).

	Euclid	Legendre	Cuthbertson	Wilson	Pierce	Willock	'Syllabus'-Manual	Chauvenet	Loomis	Morell	Reynolds	Wright
3												
4	P		P	P		P	P	P			P	
5	P	P	P				P	P	P	P	P	P
6												
7												
8												
9			P					P	P	P	P	P
10			P									
11					A							
12				A		A		Def				
13				A								

3. $ct_m \,\mathbb{P}\, C.$ 'Lines, which have a common point and make equal ∠s with a transversal, are coincidental.'

4. $X \,\mathbb{P}\, T_{m'}.$ 'Intersectional Lines make unequal ∠s with any transversal.' (N.B. This is equivalent to '$X \,\mathbb{P}\, T_{d'}$, $T_{d'}$, $T_{i'}$, $T_{f'}$, $T_{a''}$, $T_{e''}$, and $T_{i''}$': the full statements of which are as follows :—

     $X \,\mathbb{P}\, T_{d'}.$ 'Intersectional Lines make unequal alt ∠s with any transversal.'

     $X \,\mathbb{P}\, T_{d'}.$ 'Intersectional Lines make unequal ext and int opp ∠s with any transversal.'

     $X \,\mathbb{P}\, T_{i'}.$ 'Intersectional Lines make with any transversal two int ∠s ≠ two right angles.

TABLE I.   229

$X \, ℙ \, T_{\prime}.$ 'Any transversal, which is perpendicular to one of two intersectional Lines, is not so to the other.'

$X \, ℙ \, T_{a''}.$ 'Intersectional Lines make unequal alt $\angle$ s with any transversal, that which is further from their point of intersection being $>$ the other.' That is (since two intersectional Lines and a transversal constitute a triangle) 'An ext $\angle$ of a $\triangle$ $>$ the int alt $\angle$.'

$X \, ℙ \, T_{e''}.$ 'An ext $\angle$ of a $\triangle$ $>$ the int opp $\angle$.'

$X \, ℙ \, T_{i''}.$ 'Two int $\angle$ s of a $\triangle$ $<$ 2 $\llcorner$ s.'

5. $st_m \, ℙ \, S.$ 'Different Lines, which make equal $\angle$ s with a transversal, are separational.' (N.B. This is equivalent to 'st_a, or st_e, or st_i, or st_r, $ℙ$ S.')

6. $ce \, ℙ \, C.$ 'Lines, which have a common point and of which one contains 2 points on the same side of the other and equidistant from it, are coincidental.'

7. $X \, ℙ \, F.$ 'Intersectional Lines are such that all points on each, which lie on the same side of the other, are unequally distant from it.'

8. $se \, ℙ \, S.$ 'Different Lines, of which one contains 2 points on the same side of the other and equidistant from it, are separational.'

9. sS real.   'If there be given a Line and a point not on it; a Line can be drawn, through the given point, and separational from the given Line.

10. $X \, ℙ \, V.$ 'Intersectional Lines diverge without limit; i.e. a point can be found on each, whose distance from the other shall exceed any assigned length.'

11. $cD \, ℙ \, C.$ 'Lines, which have a common point and the same direction, are coincidental.'

12. $X \, ℙ \, D'.$ 'Intersectional Lines have different directions.'

13. sD' real.   'If there be given a Line and a point not on it; a Line can be drawn, through the given point, and having a direction different from that of the given Line.'

TABLE I (*continued*).

	Euclid	Legendre	Cuthbertson	Wilson	Pierce	Willock	'Syllabus'-Manual	Chauvenet	Loomis	Morell	Reynolds	Wright
14												
15 (1)						A						
(2)												
16												
17						A						
18												
19			A									

14. *ce* $\mathbb{P}$ *D*.   'Lines, which have a common point and of which one contains 2 points on the same side of the other and equidistant from it, have the same direction.'

15 (1). *D'* $\mathbb{P}$ S_{-1}.   'Lines, which have different directions, have all points, but one, separate.' (Nothing is asserted of the excepted point.)

15 (2). *c D'* $\mathbb{P}$ *X*.   'Lines, which have a common point and different directions, are intersectional.'

16. $\overline{C\,|\,c\,|}$ $\mathbb{P}$ $\overline{\quad|\quad|\,c}$.   'A Line, which has a point common with one of two coincidental Lines, has a point common with the other also.'

17. $\overline{c\,|\,c\,|S}$ $\mathbb{P}$ $\overline{X\,|\quad|\quad}$; $\equiv$ $\mathbb{P}$ $\overline{X\,|\,X\,|}$.   'A Line, which has a point common with one of two separational Lines, and also a point common with the other, intersects the one.' This $\equiv$ 'it intersects both.'

TABLE I.   231

18. $\overline{\;|\,c\,|\,S\;}$ $\mathbb{P}$ $\overline{s\,|\;}$ $|$.   'A Line, which has a point common with one of two separational Lines, has a point separate from the other.'

19. $\overline{\,sE\,|\,c\,V\,|\,}$ $\mathbb{P}$ $\overline{\;\;|\;\;}$ $|\,\overline{X}$.   'If, of two different Lines, the second (*not* assumed to be straight,) be equidistant from the first; a third Line, which has a point common with, and diverges without limit from the first, is intersectional with the second.'

TABLE II. [See p. 35.]

	Euclid	Legendre	Cuthbertson	Wilson	Pierce	Willock	'Syllabus-Manual	Chauvenet	Loomis	Morell	Reynolds	Wright
1 (1)		P	P					P	P	P		P
(2)	P	P	P				P	P	P	P	P	P
2 (1)												
(2)*	A	P				P		P	P	P		
3												
4							P					
5												
6		P	P					P	P	P		P
7*								P				
8*												
9												
10												
11			P									
12												

1 (1). 'Angles, which have the sides of the one separational from those of the other, each from each, are equal.'

1 (2). $S \mathbb{P} T_m$.   'Separational Lines make equal $\angle$ s with any transversal.'

2 (1). $t_{m'} \mathbb{P} c$.   'Lines, which make unequal $\angle$ s with a transversal, have a common point.'

2 (2). $st_{m'} \mathbb{P} X$.   'Different Lines, which make unequal $\angle$ s with a transversal, are intersectional.' (N.B. This includes the following as a particular case.)

TABLE II. 233

Euclid's Axiom.

$st_{i'}$ ℙ $X.$ '*Different Lines, which make with a transversal* 2 *int* ∠s < 2 ∟ s, *are intersectional.*'

3. sT_m real. 'If there be given a Line and a point not on it; a Line can be drawn, through the given point, and such that it and the given Line make equal ∠s with any transversal.'

4. st_m ℙ $T_m.$ 'Different Lines, which make equal ∠s with a certain transversal, do so with any transversal.'

5. $st_{m'}$ ℙ $T_{m'}.$ 'Different Lines, which make unequal ∠s with a certain transversal, do so with any transversal.'

6. S ℙ $E.$ 'Separational Lines are equidistant.'

Simpson's Axiom.

7. sf ℙ $X.$ '*Different Lines, of which one contains* 2 *points unequally distant from the other, are intersectional.*'

8. sE real. 'If there be given a Line and a point not on it; a Line can be drawn, through the given point, and such that it and the given Line shall be equidistant from each other.' (N.B. This includes the following as a particular case.)

Clavius' Axiom.

'If there be given a Line and a point not on it; a Line can be drawn, through the given point, equidistant from the given Line.

9. se ℙ $T_m.$ 'Different Lines, of which one has 2 points on the same side of the other and equidistant from it, make equal ∠s with any transversal.'

10. $st_{m'}$ ℙ $F.$ 'Different Lines, which make unequal ∠s with a transversal, are such that all points on each, which lie on the same side of the other, are unequally distant from it.'

11. st_m ℙ $E.$ 'Different Lines, which make equal ∠s with a transversal, are equidistant.'

12. sf ℙ $T_{m'}.$ 'Different Lines, of which one has 2 points unequally distant from the other, make unequal ∠s with any transversal.'

TABLE II (*continued*).

	Euclid	Legendre	Cuthbertson	Wilson	Pierce	Willock	'Syllabus'-Manual	Chauvenet	Loomis	Morell	Reynolds	Wright
{ 13 { 14												
{ 15(1)												
(2)			P									
(3)	P	P	P				P	A	P	A	A	A
16(1)*							A	P	A	P	P	P
(2)								A		P	A	P
17*			A									
18(1)	P	P	P	P	P	P	P	P	P	P	P	P
(2)	P	P	P	P	P	P	P	P	P	P	P	P

13. *se* ⫪ *E.*   'Different Lines, of which one has 2 points on the same side of the other and equidistant from it, are equidistant from each other.'

14. *sf* ⫪ *F.*   'Different Lines, of which one has 2 points unequally distant from the other, are such that all points on each, which lie on the same side of the other, are unequally distant from it.'

15 (1). $\overline{\;|S|S\; ⫪\; \overline{X'}|\;\;|\;}$.   'Lines, which are separational from a third Line, are not intersectional.'

15 (2). $\overline{\;c|S|S\; ⫪\; \overline{C}|\;\;|\;}$; ≡ *sS* unique.   'Lines, which have a common point and are separational from a third Line, are coincidental.' This ≡ 'If there be given a Line and

TABLE II.   235

a point not on it; only one Line can be drawn, through the given point, separational from the given Line.'

15 (3). $\overline{s\,|\,S\,|\,S}$ $\mathbb{P}\,\overline{S\,|}\ \ |$.   'Different Lines, which are separational from a third Line, are separational from each other.'

Playfair's Axiom.

16 (1). $\overline{X\,|\ \ |}$ $\mathbb{P}\overline{(\ \ |\,S\,|\,S}\,)'$.   '*Intersectional Lines cannot both be separational from a third Line.*'

16 (2). $\overline{X\,|\,S\,|}$ $\mathbb{P}\ \overline{|\ \ |\,X}$.   'A Line, which is in-tersectional with one of two separational Lines, is intersectional with the other also.'

The Ahmedabad Axiom.

17. '*A Line cannot recede from and then approach another; nor can one approach and then recede from another on the same side of it.*'

18 (1). 'An ext $\angle$ of a $\Delta = 2$ int opp $\angle$ s.'

18 (2). 'The 3 $\angle$ s of a $\Delta = 2 \llcorner$ s.'

TABLE III. [See p. 40.]

	WILSON	PIERCE	WILLOCK
1	P	P	P
2 (1)	P	P	
(2)	P	P	P
3	P	P	
4			
5 (1)			
(2)			P
6 (1)			
(2)			
(3)	P	P	A
7 (1)			
(2)		P	P.
(3)			
8			P

1. $sD \, \mathbb{P} \, S.$ ''sD'- related Lines are separational.'

2 (1). 'Angles, which have the sides of the one 'sD'- related to those of the other, each to each, are equal.

2 (2). $sD \, \mathbb{P} \, T_m.$ ''sD'-related Lines make equal $\angle$ s with any transversal.'

3. $sD \, \mathbb{P} \, E.$ ''sD'-related Lines are equidistant.'

4. $\overline{\;|\; C \;|sD\;| \mathbb{P} \;\overline{S|}\; |\;}$.   'A Line, which is coincidental with one of two 'sD'-related Lines, is separational from the other.'

5 (1). $c \;|\quad |sD \;|\; \mathbb{P} \;\overline{\;|s|}\;$.   'A Line, which has a point common with one of two 'sD'-related Lines, has a point separate from the other.'

TABLE III. 237

5 (2). $\quad c \mid c \mid \overline{sD}$ ℙ $\overline{\mid X \mid}$. 'A Line, which has a point common with one of two 'sD'-related Lines and also a point common with the other, is intersectional with the one.' This $\equiv$ 'it is intersectional with both.'

6 (1). $\quad \overline{\mid sD \mid sD}$ ℙ $\overline{D \mid \quad \mid}$. 'Lines, which are 'sD'-related to a third Line, have the same direction.'

6 (2). $\quad c \mid \overline{sD \mid sD}$ ℙ $\overline{cD \mid \quad \mid}$. 'Lines, which have a common point and are 'sD'-related to a third Line, have a common point and the same direction.'

6 (3). $\quad s \mid \overline{sD \mid sD}$ ℙ $\overline{sD \mid \quad \mid}$. 'Different Lines, which are 'sD'-related to a third Line, are 'sD'-related to each other.'

7 (1). $\quad \overline{\mid sD \mid sD}$ ℙ $\overline{X' \mid \quad \mid}$. 'Lines, which are 'sD'-related to a third Line, are not intersectional.'

7 (2). $\quad c \mid \overline{sD \mid sD}$ ℙ $\overline{C \mid \quad \mid}$. 'Lines, which have a common point and are 'sD'-related to a third Line, are coincidental.' This $\equiv$ 'If there be given a Line and a point not on it; only one Line can be drawn, through the given point, and 'sD'-related to the given Line.'

7 (3). $\quad s \mid \overline{sD \mid sD}$ ℙ $\overline{S \mid \quad \mid}$. 'Different Lines, which are 'sD'- related to a third Line, are separational.'

8. $\quad \overline{X \mid \quad \mid}$ ℙ $\overline{(\quad \mid sD \mid sD)'}$. 'Intersectional Lines cannot both be 'sD'-related to a third Line.'

TABLE IV. [See p. 41.]

	Wilson	Pierce	Willock
1	A	A	A
{ 2			A
3 (1)			
(2)	A		
4	A		
{ 5	P	P	P
{ 6			
{ 7	A		
{ 8			

1. *sD* real. 'If there be given a Line and a point not on it; a Line can be drawn, through the given point, and '*sD*'- related to the given Line.'

2. *S* ℙ *D.* 'Separational Lines have the same direction.'

3 (1). *D'* ℙ *c.* 'Lines, which have different directions, have a common point.'

3 (2). *sD'* ℙ *X.* 'Different Lines, which have different directions, are intersectional.'

4. 'An angle may be transferred, preserving its magnitude and the directions of its sides.' This ≡ 'If there be given a Line and a point not on it; a Line can be drawn, through the given point, '*sD*'-related to the given Line, and such that it and the given Line make equal ∠s with any transversal.'

5. st_m ℙ *D.* 'Different Lines, which make equal ∠s with a transversal, have the same direction.'

6. *sD'* ℙ $T_{m'}$. 'Different Lines, which have different directions, make unequal ∠s with any transversal.'

TABLE IV. 239

7. *se* ℙ *D.*   'Different Lines, of which one contains 2 points on the same side of the other and equidistant from it, have the same direction.'

8. *sD'* ℙ *F.*   'Different Lines, which have different directions, are such that all points on each, which lie on the same side of the other, are unequally distant from it.'

APPENDIX IV.

The various methods of treating Parallels, adopted by Euclid and his Modern Rivals.

[N.B.—In each Section, the first Table gives the logical genealogy of the propositions assumed or proved by the writer, numbered in the order in which they occur, as well as their enunciations, expressed symbolically, so far as that is possible; the second Table gives, first, references to the genealogical Table, secondly references to Tables I–IV, thirdly the enunciations in full, and fourthly references to the writer in question.]

§ 1. EUCLID.

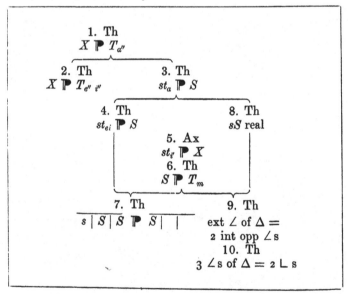

§ 1. Euclid.

	Tables I–IV		Euclid.
1. Th	I. pt of 4	, Ext $\angle$ of Δ > int alt $\angle$	I. pt of 16
2. ,,	rem of ,,	Ext $\angle$ of Δ > int opp $\angle$; and 2 int $\angle$ s of Δ < 2 $\llcorner$ s	I. rem of 16, and 17
3. ,,	I. pt of 5	Different Lines, which make with a transversal equal alt $\angle$ s, are separational	I. 27
4. ,,	rem of ,,	Different Lines, which make with a transversal equal ext and int $\angle$ s or 2 int $\angle$ s = 2 $\llcorner$ s, are separational	28
5. Ax	II. 2 pt of (2)	Different Lines, which make with a transversal 2 int $\angle$ s < 2 $\llcorner$ s, are intersectional	Ax. 12
6. Th	II. 1 (2)	Separational Lines make, with any transversal, equal alt $\angle$ s, equal ext and int $\angle$ s, and 2 int $\angle$ s = 2 $\llcorner$ s	I. 29
7. ,,	15 (3)	Different Lines, which are separational from the same Line, are separational from each other	30
8. ,,	I. 9	If there be given a Line and a point not on it; a Line can be drawn, through the given point, separational from the given Line	31. Not stated, but virtually contained in Problem
9. .,	18 (1)	Ext $\angle$ of Δ = 2 int opp $\angle$ s	pt of 32
10. .,	(2)	3 int $\angle$ s of Δ = 2 $\llcorner$ s	rem of 32

§ 2.　Legendre.

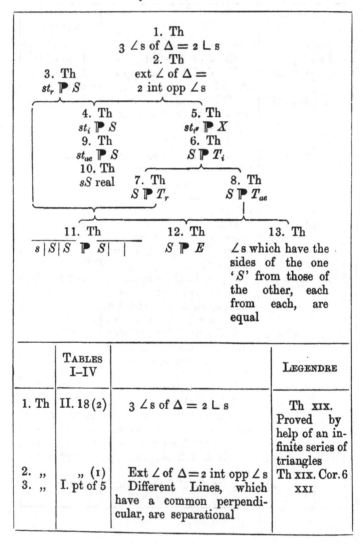

1. Th

3 ∠s of △ = 2 ∟ s

2. Th

ext ∠ of △ =
2 int opp ∠s

3. Th
st_r ℙ S

4. Th
st_i ℙ S

5. Th
$st_{i'}$ ℙ X

9. Th
st_{ae} ℙ S

6. Th
S ℙ T_i

10. Th
sS real

7. Th
S ℙ T_r

8. Th
S ℙ T_{ae}

11. Th

$s \,|\, S \,|\, S$　ℙ　$\overline{S}|$　|

12. Th

S ℙ E

13. Th

∠s which have the
sides of the one
'S' from those of
the other, each
from each, are
equal

	TABLES I–IV		LEGENDRE
1. Th	II. 18 (2)	3 ∠s of △ = 2 ∟ s	Th xix. Proved by help of an infinite series of triangles
2. ,,	,, (1)	Ext ∠ of △ = 2 int opp ∠ s	Th xix. Cor. 6
3. ,,	I. pt of 5	Different Lines, which have a common perpendicular, are separational	xxi

§ 2. LEGENDRE.

	TABLES I-IV		LEGENDRE
4. Th	I. pt of 5	Different Lines, which make with a transversal 2 int ∠s = 2 ∟s, are separational	Th XXII
5. „	II. 2 pt of (2)	Different Lines, which make with a transversal 2 int ∟s < 2 ∟s, are intersectional	XXIII. From (1), by an infinite series of angles
6. „	II. 1 pt of (2)	Separational Lines make with any transversal 2 int ∠s = 2 ∟s	Th XXIV: Contranominal of (5)
7. „	„	Any transversal, which is ⊥ one of two separational Lines, is ⊥ the other	do. Cor. 1
8. „	II. 1 rem of (2)	Separational Lines make with any transversal equal alt ∠s and equal ext and int ∠s	do. Cor. 2
9. „	I. rem of 5	Different Lines, which make with a transversal equal alt ∠s or equal ext and int ∠s, are separational	do. Scholium
10. „	I. 9	If there be given a Line and a point not on it; a Line can be drawn, through the given point, separational from the given Line	Pr. VI. Not stated, but virtually contained in Problem.
11. „	II. 15 (3)	Lines, which are separational from the same Line, are separational from each other	Th XXV
12. „	6	Separational Lines are equidistant	XXVI
13. „	1 (1)	Angles, which have the sides of the one separational from those of the other, each from each, are equal	XXVII

§ 3. Cuthbertson.

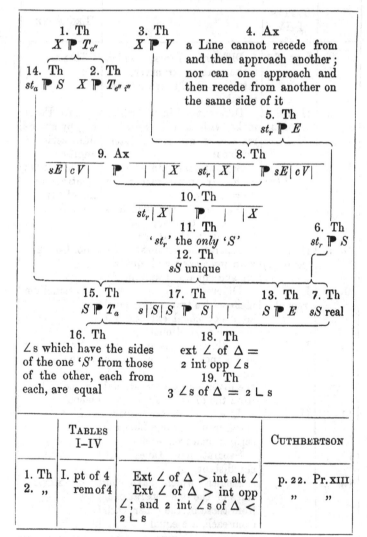

	TABLES I–IV		CUTHBERTSON
1. Th	I. pt of 4	Ext ∠ of Δ > int alt ∠	p. 22. Pr. XIII
2. „	rem of 4	Ext ∠ of Δ > int opp ∠; and 2 int ∠s of Δ < 2 ∟ s	„ „

§ 3. CUTHBERTSON.

	TABLES I–IV		CUTHBERTSON
3. Th	I. 10	Intersectional Lines diverge without limit; i.e. a point can be found, on either of them, whose distance from the other shall exceed any assigned length	p. 33. Ded. G
4. Ax	II. 17	A Line cannot recede from and then approach another; nor can one approach and then recede from another on the same side of it	p. 34. Ax
5. Th	II. pt of 11	Different Lines, which have a common perpendicular, are equidistant	p. 35. l. 5. '∴ all points &c.' down to 'equidistant from RS.' Proved by making diagram revolve
6. „	I. pt of 5	Different Lines, which have a common perpendicular, are separational	p. 35. Alluded to in the words 'no other line *besides RAS* &c.'
7. „	I. 9	A Line can be drawn, separational from a given Line, through a given point not on it	p. 34. Occurs in enunciation; also alluded to in the above words
8. „	not in Tab	(Asserts 3 and 5)	p. 35. Necessary link in argument : virtually asserted in the words 'Again &c.'

§ 3. Cuthbertson.

	Tables I–IV		Cuthbertson
9. Ax	I. 19	Two Lines, which are such that a third Line is equidistant from one of them and diverges without limit from the other, are intersectional	p. 35. Tacitly assumed in the words '.˙. the straight line &c.'
10. Th	not in Tab	Two Lines, which are such that a third Line has a common perpendicular with one of them and is intersectional with the other, are intersectional	p. 35. Virtually asserted in the above words
11. „	„	If there be given a Line and a point not on it : the Line, drawn through the point so as to have a common perpendicular with the given Line, is the *only* one which can be drawn, through the point, separational from the given Line	p. 35. 'Hence through *A* &c.'
12. „	II. 15 (2)	Only one Line can be drawn, separational from a given Line, through a given point not on it	p. 34, 35. Asserted in enunciation, and implied as *general* conclusion
13. „	6	Separational Lines are equidistant	p. 34, 35. Asserted in enunciation, and deducible from argument
14. „	I. 5	Different Lines, which make equal angles with a transversal, are separational	p. 36. Pr. xx

§ 3. Cuthbertson.

	Tables I–IV		Cuthbertson
15. Th	II. 1 (2)	Separational Lines make equal angles with any transversal	p. 38. Pr. xxi
16. „	„ (1)	Angles, which have the sides of the one separational from those of the other, each from each, are equal	p. 39. Pr. xxi. Cor.
17. „	15 (3)	Different Lines, which are separational from the same Line, are separational from each other	p. 40. Pr. xxii
18. „	18 (1)	Ext ∠ of Δ = 2 int opp ∠s	p. 42. Pr. xxiv
19. „	„ (2)	3 ∠s of Δ = 2 ∟s	„ „

§ 4. Wilson.

1. Ax 2. Ax 3. Ax 7. Ax

sD real sD' real $X \mathrel{\mathbb{P}} D'$ an $\angle$ may be transferred, preserving its magnitude and the directions of its sides

4. Ax

$sD' \mathrel{\mathbb{P}} X$

5. Th 10. Th 8. Th

$sD \mathrel{\mathbb{P}} S$ $X \mathrel{\mathbb{P}} T_{e'}$ $\angle$ s, which have the

11. Th sides of the one

$(sD)' \mathrel{\mathbb{P}} T_{d'i'}$ 'sD' with those of

12. Th the other, each with

$st_m \mathrel{\mathbb{P}} D$ each, are equal

6. Th 9. Th

$s \mid sD \mid sD \mid \mathrel{\mathbb{P}} sD \mid \quad \mid$ $sD \mathrel{\mathbb{P}} T_e$

13. Th 16. Th

ext $\angle$ of $\Delta = $ int opp $\angle$ s $sD \mathrel{\mathbb{P}} E$

14. Th

$3 \angle$ s of $\Delta = 2 \mathrel{L}$ s

17. Ax 15. Th

$se \mathrel{\mathbb{P}} D$ $X \mathrel{\mathbb{P}} T_{d''e''}$

	Tables I–IV		Wilson
1. Ax	IV. 1	If there be given a Line and a point not on it, a Line can be drawn, through the given point, and 'sD'-related to the given Line	p. 11 Ax. 6.
2. ,,	I. 13	If there be given a Line and a point not on it, a Line can be drawn through the given point, and having a direction different from that of the given Line	,, ,,

§ 4. Wilson.

	Tables I–IV		Wilson
3. Ax	I. 12	Intersectional Lines have different directions	p. 11. Ax. 7
4. „	IV. 3 (2)	Different Lines, which have different directions, are intersectional	„ Ax. 8
5. Th	III. 1	'*sD*'-related Lines are separational	„ Result (1)
6.	„ 6 (3)	Different Lines, which are '*sD*'-related to the same Line, are '*sD*'-related to each other	„ Result (2). Proved as theorem from nature of direction: but quite axiomatic, though ideal
7. Ax	IV. 4	An ∠ may be transferred, preserving its magnitude and the directions of its sides	p. 12. Ax. 9
8. Th	III. 2 (1)	Angles, which have the sides of the one '*sD*'-related to those of the other, each to each, are equal	„ Th. 4. Merely another form of the above axiom
9. Th	2. pt of (2)	'*sD*'-related Lines make equal ext and int opp ∠s with any transversal	p. 12. Th. 4. Cor.
10. Th	I. 4	Intersectional Lines make unequal ext and int opp ∠s with any transversal	p. 14. Th. 5
11. Th	not in Tab	Lines, which are *not* '*sD*'-related, make, with any transversal, unequal alt ∠s and also 2 int ∠s ≠ 2 ∟s	p. 14. Th. 5. Cor. 1. *False.* The *correct* form would be 'Different Lines, which have different directions, make &c.'

§ 4. Wilson.

	Tables I–IV		Wilson
12. Th	IV. 5	Different Lines, which make equal $\angle$s with a transversal, have the same direction	p. 20. Cor. 2. Proved from Cor. 1. As it happens, it follows equally well from the *correct* form of Cor. 1.
13. „	II. 18 (1)	Ext $\angle$ of $\Delta = 2$ int opp $\angle$s	p. 20. Th. 7
14. „	(2)	3 $\angle$s of $\Delta = 2 \llcorner$s	„ „
15. „	I. pt of 4	Ext $\angle$ of $\Delta >$ each int opp $\angle$	„ „ Cor. 4
16. „	III. 3	'sD'-related Lines are equidistant	p. 38. Exercise 9
17. Ax	IV. 7	Different Lines, of which one has two points on the same side of the other and equidistant from it, have the same direction	p. 59. Assumed in course of Th. 24, in the words 'that is, CD will be parallel to AB'

[Turn over.

§ 5. Pierce.

$$1.\ \text{Ax}$$
$$cD\ ℗\ C$$
$$2.\ \text{Th}$$
$$\overline{c\ |sD|sD\ ℗\ C|\quad|}$$
$$3.\ \text{Th}$$
$$sD\ ℗\ S$$

4. Th
∠s, which have the sides of the one 'sD' with those of the other, are equal

5. Th
$$sD\ ℗\ T_{ae}$$

6. Ax
sD real

7. Th
$$st_{ae}\ ℗\ D$$

8. Th
$$sD\ ℗\ T_i$$
9. Th
$$st_i\ ℗\ D$$

10. Th
$$sD\ ℗\ T_r$$
11. Th
$$st_r\ ℗\ D$$

13. Th
$$3\ ∠\text{s of } Δ = 2 ∟\text{s}$$
14. Th
$$\text{ext } ∠ \text{ of } Δ = 2$$
$$\text{int opp } ∠\text{s}$$

12. Th
$$\overline{s\ |sD|sD\ ℗\ sD|\quad|}$$

15. Th
$$sD\ ℗\ E$$

	Tables I–IV		Pierce
1. Ax	I. 11	Lines, which have a common point and the same direction, are coincidental	p. 9. § 28. Assumed in course of argument
2. Th	III. 7 (2)	Lines, which have a common point, and which are 'sD'-related to a third Line, are coincidental	p. 9. § 28. Proved in course of argument
3. „	III. 1	'sD'-related Lines are separational	p. 9. § 28
4. „	„ 2 (1)	Angles, which have the sides of the one 'sD'-related to those of the other, each to each, are equal	„ § 29
5. „	„ 2. pt of (2)	'sD'-related Lines make with any transversal equal alt ∠s and equal ext and int opp ∠s	„ § 30

§ 5. Pierce.

	Tables I–IV		Pierce
6. Ax	IV. 1	If there be given a Line and a point not on it; a Line can be drawn, through the given point, and '*sD*'-related to the given Line	p. 10. § 31. Assumed in course of argument
7. Th	,, pt of 5	Different Lines, which make with a transversal equal alt ∠s or equal ext and int opp ∠s, have the same direction	p. 10. § 31
8. ,,	III 2. rem of (2)	'*sD*'-related Lines make with any transversal 2 int ∠s = 2 ∟ s	,, § 32
9. ,,	IV. rem of 5	Different Lines, which make with a transversal 2 int ∠s = 2 ∟ s, have the same direction	,, § 33
10. ,,	III. 2. pt of (2)	A Line, which is ⊥ one of two '*sD*'-related Lines, is ⊥ the other	,, § 34. Superfluous: included in 5
11. ,,	IV. pt of 5	Different Lines, which have a common perpendicular, have the same direction	p. 11. § 35. Superfluous: included in 7
12. ,,	III. 6 (3)	Different Lines, which are '*sD*'-related to a third Line, have the same direction	p. 11. § 36. Proved from nature of direction, but quite axiomatic
13. ,,	II. 18 (2)	3 ∠s of a △ = 2 ∟ s	p. 20. § 65
14. ,,	,, (1)	An ext ∠ of a △ = 2 int opp ∠s	p. 21. § 71
15. ,,	III. 3	'*sD*'-related Lines are equidistant	p. 23. § 82

§ 6. Willock.

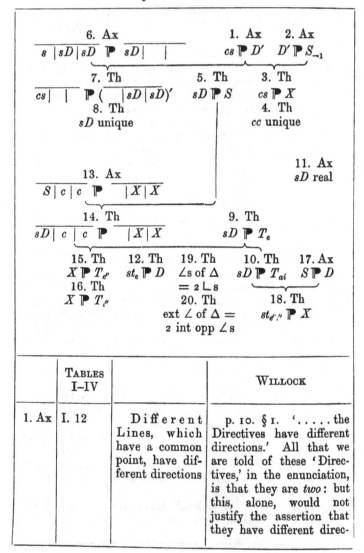

	Tables I–IV		Willock
1. Ax	I. 12	Different Lines, which have a common point, have different directions	p. 10. § 1. '.....the Directives have different directions.' All that we are told of these 'Directives,' in the enunciation, is that they are *two*: but this, alone, would not justify the assertion that they have different direc-

§ 6. Willock.

	Tables I–IV		Willock
			tions, for they might be parallels. The enunciation is clearly meant to be ' Two Directives, *which have a common point,* can &c '
2. Ax	I. 15(1)	Lines, which have different directions, have all points, but one, separate	p. 10. § 1. ' since the Directives have different directions, they must . . . . not meet again.'
3. Th	I. 2(2)	D i f f e r e n t L i n e s, w h i c h have a common point, are inter-sectional	p. 10. § 1.
4. „	„ 2 (1)	Through two given points only one Line can be drawn	p. 10. § 1. ' This is equivalent to Euclid's 10th Axiom—" two right Lines cannot enclose a space." ' This is not quite correct. The 10th Axiom asserts '$s \, \mathbb{P} \, S_{-1}$'; while this article asserts '$cs \, \mathbb{P} \, X$'; so that its subject has an additional property 'c'.
5. „	III. 1.	'sD'-related Lines are sepa-rational	p. 11. § 5.
6. Ax	„ 6 (3)'	D i f f e r e n t Lines, which are 'sD'-related to the same Line, are 'sD'-related to each other	p. 12. § 7. ' if two ' (parallels) ' could be drawn, each would have the same direction as that of the directive ' (and so the same direction as each other)

§ 6. Willock.

	Tables I–IV		Willock
7. Th	III. 8	Intersectional Lines cannot both be '*sD*'-related to the same Line	p. 12. § 7. 'these two parallels' (which therefore are '*s*') 'pass through the point' (and thus are '*c*'). 'They must therefore . . . . not be both parallels to the directive.'
8. ,,	,, 7(2)	If there be given a Line and a point not on it ; only one Line can be drawn, through the given point, and '*sD*'-related to the given Line	p. 12. § 7. This must not be taken to assert that 'one and only one can be drawn' but merely to deny that *more* than one can be drawn
9. ,,	,, 2. pt of (2)	'*sD*'-related Lines make with any transversal equal ext and int opp ∠s	p. 12. § 8. Proved from nature of direction, an 'angle' being regarded as 'the difference of two directions'
10. ,,	,, 2. rem.of(2)	'*sD*'-related Lines make with any transversal equal alt ∠s and 2 int ∠s=2 ∟s	p. 12. § 9
11. Ax	IV. 1	If there be given a Line and a point not on it ; a Line can be drawn, through the given point, and '*sD*'-related to the given Line	p. 13. § 10. 'some . . . line. . . . . through O, is parallel to X.'

§ 6. WILLOCK.

	TABLES I–IV		WILLOCK
12. Th	IV. pt of 5	D i ff e r e n t Lines, which make equal ext and int opp ∠s with a transversal, have the same direction	p. 13. § 10
13. Ax	I. 17	A Line, which has a point common with one of two separational Lines, and also a point common with the other, is intersectional with both	p. 14. § 12. 'the angle ZOX, *which is greater than* a_1.' Why greater? Clearly he assumes that X must diverge from (i. e. intersect) Y_1. (His extreme parsimony of letters in this diagram is most confusing.) And why must it diverge? Clearly because, starting from a point *on* Y_1, it (intersects Y, i.e.) reaches a point on a Line parallel to Y_1, i. e. it reaches a point *not* on Y_1. But why should a point, on a Line parallel to Y_1, be necessarily *not* on Y_1? Because parallel Lines are separational. Hence it is clear that he assumes it as axiomatic that Y and Y_1 being separational, and X passing through a point on each, X must intersect Y_1
14. Th	III. 5 (2)	A Line, which has a point common with one of	p. 14. § 12. From the above (tacitly assumed) axiom, be clearly intends to

§ 6. Willock.

	Tables I–IV		Willock
		two '*sD*'-related Lines, and also a point common with the other, is intersectional with both	draw the conclusion that the fact, of *Y* and *Y₁* being '*sD*'-related, leads to the same result
15. Th	I. pt of 4	An ext ∠ of a Δ < int opp ∠	p. 14. § 12. Note that he omits the other part of Euc. I. 16, that it is > the int *alt* ∠
16. „	I. pt of 4	2 int ∠s of a Δ < 2 ∟ s	p. 14. § 12
17. Ax	IV. 2	Separational Lines have the same direction	„ § 13. 'suppose they do not meet. Then they should be parallel'
18. Th	II. 2. pt of (2)	Different Lines, which make with a transversal an ext ∠ > the int opp ∠, or 2 int ∠s < 2 ∟ s, are intersectional	p. 14. § 13
19. „	„ 18 (2)	3 ∠s of a Δ = 2 ∟ s	„ § 14
20. „	„ (1)	An ext ∠ of a Δ = 2 int opp ∠s	p. 15. § 14. Cor. 1

[Turn over.

§ 7. CHAUVENET.

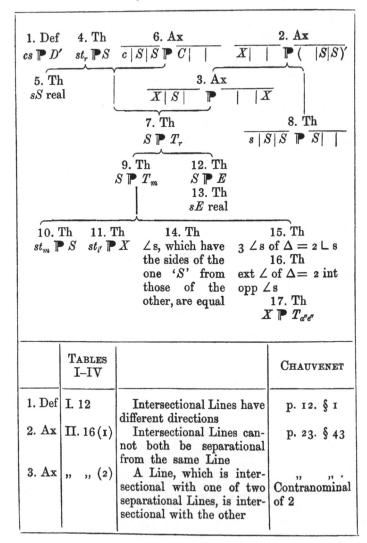

	TABLES I–IV		CHAUVENET
1. Def	I. 12	Intersectional Lines have different directions	p. 12. § 1
2. Ax	II. 16 (1)	Intersectional Lines cannot both be separational from the same Line	p. 23. § 43
3. Ax	,, ,, (2)	A Line, which is intersectional with one of two separational Lines, is intersectional with the other	,, ,, . Contranominal of 2

§ 7. Chauvenet.

	Tables I–IV		Chauvenet
4. Th	I. pt of 5	Different Lines, which have a common perpendicular, are separational	p. 23. Pr. XI. proved from earlier geometry
5. Th	„ 9	If there be given a Line and a point not on it; a Line can be drawn, through the given point, and separational from the given Line	p. 23. Pr. XI. Cor. 1
6. Ax	II. 15(2)	˙ Lines, which have a common point and are separational from the same Line, are coincidental	p. 23. Cor. 2. ' Moreover the perpendicular . . . is parallel to AB . . . . and must coincide with CD.'
7. Th	„ 1. pt of (2)	A Line, which is ⊥ one of two separational Lines, is ⊥ the other	p. 23. § 46
8. „	„ 15 (3)	Different Lines, which are separational from the same Line, are separational from each other	p. 24. Pr. XII. Contranominal of 2
9. Th	II. 1 (2)	Separational Lines make equal ∠s with any transversal	p. 24. Pr. XIII, and Cors
10. „	I. 5	Different Lines, which make equal ∠s with a transversal, are separational	p. 25. Pr. XIV, and Cors 1 and 2
11. „	II. 2. pt of (2)	Different Lines, which make with a transversal 2 int ∠s $\neq$ 2 ∟ s, are intersectional	p. 26. Pr. XIV. Cor. 3. Incorrectly deduced from 9 and 10, the latter being superfluous

§ 7. Chauvenet.

	Tables I–IV		Chauvenet
12. Th	„ 6	Separational Lines are equidistant	p. 26. Pr. xv
13. „	„ 8	If there be given a Line and a point not on it; a Line can be drawn through the given point, such that the two Lines shall be equidistant from each other	„ „ Cor
14. „	II. 1 (1)	Angles, which have the sides of the one separational from those of the other, are equal	p. 27. Pr. xvi
15. „	„ 18 (2)	3 ∠s of a Δ = 2 ∟ s	p. 29. Pr. xviii
16. „	„ (1)	An ext ∠ of a Δ = 2 int opp ∠s	„ „ Cor. 1
17. „	I. pt of 4	An ext ∠ of a Δ > either of the int opp ∠s	„ „ „

[Turn over.

§ 8. Loomis.

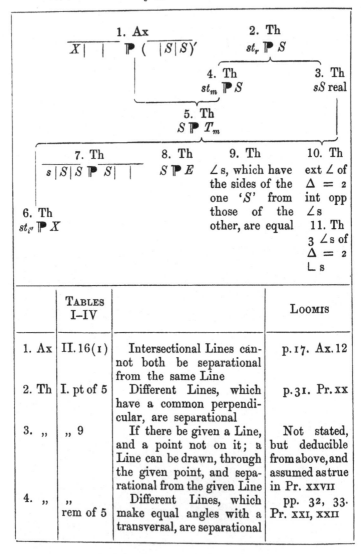

	Tables I–IV		Loomis
1. Ax	II. 16 (1)	Intersectional Lines cannot both be separational from the same Line	p. 17. Ax. 12
2. Th	I. pt of 5	Different Lines, which have a common perpendicular, are separational	p. 31. Pr. xx
3. „	„ 9	If there be given a Line, and a point not on it; a Line can be drawn, through the given point, and separational from the given Line	Not stated, but deducible from above, and assumed as true in Pr. xxvii
4. „	„ rem of 5	Different Lines, which make equal angles with a transversal, are separational	pp. 32, 33. Pr. xxi, xxii

§ 8. Loomis.

	Tables I–IV		Loomis
5. Th	II. 1 (2)	Separational Lines make equal $\angle$ s with any transversal	p. 33. Pr. XXIII
6. ,,	,, 2 pt of (2)	Different Lines, which make with a transversal 2 int $\angle$ s $<$ 2 $\llcorner$ s, are intersectional	,, ,, Cor. 3
7. ,,	,, 15 (3)	Different Lines, which are separational from a third Line, are separational from each other	p. 34. Pr. XXIV
8. ,,	,, 6	Separational Lines are equidistant	p. 35. Pr. XXV
9. ,,	,, 1 (1)	Angles, which have the sides of the one separational from those of the other, each from each, are equal	,, Pr. XXVI
10. ,,	,, 18 (1)	An ext $\angle$ of a Δ = 2 int opp $\angle$ s	p. 36. Pr. XXVII
11. ,,	,, (2)	The 3 $\angle$ s of a Δ = 2 $\llcorner$ s	,, ,,

§ 9. Morell.

1. Th st_r ℙ S	3. Ax sS unique

| 2. Th
sS real | 5. Th
S ℙ T_r
7. Th
S ℙ T_m | 4. Th
$\overline{X \mid S \mid}$ ℙ $\overline{\mid X}$ | 6. Th
$s \mid S \mid S$ ℙ $\overline{S \mid} \mid$ |

8. Th st_m ℙ S	9. Th $st_{i''}$ ℙ X	10. Th ∠s, which have the sides of the one 'S' from those of the other, are equal
11. Th $3 ∠s$ of $\triangle = 2 ∟ s$ 12. Th ext $∠$ of $\triangle =$ 2 int opp $∠s$	13. Th S ℙ E	

	Tables I–IV		Morell
1. Th	I. pt of 5	Different Lines, which have a common perpendicular, are separational	p. 21. Th. xix
2. „	I. 9	If there be given a Line and a point not on it; a Line can be drawn, through the given point, and separational from the given Line	„ Th. xx
3. Ax	II. 15(2)	If there be given a Line and a point not on it; only one Line can be drawn, through the given point, and separational from the given Line	„ Th. xx. 'and it may be admitted &c.'
4. Th	„ 16 (2)	A Line, which is intersectional with one of two separational Lines, is intersectional with the other	p. 21. Th. xx. Cor

§ 9. Morell.

	Tables I–IV		Morell
5. Th.	II. 1. pt of (2)	A Line, which is ⊥ one of two separational Lines, is ⊥ the other	p. 22. Th. xxi
6. „	„ 15 (3)	Different Lines, which are separational from a third Line, are separational from each other	„ Th. xxii
7. „	„ 1 rem of (2)	Separational Lines make equal ∠s with any transversal	p. 22–24. Th. xxiii, xxiv
8. „	I. rem of 5	Different Lines, which make equal ∠s with a transversal, are separational	p. 25. Th. xxv
9. „	II. 2. pt of (2)	Different Lines, which make with a transversal 2 int ∠s < 2 ∟s, are intersectional	p. 26. Th. xxv. Cor
10. „	„ 1 (1)	Angles, which have the sides of the one separational from those of the other, each from each, are equal	„ Th. xxvi.
11. „	„ 18 (2)	The 3 ∠s of a △ = 2 ∟s	p. 27. Th. xxviii
12. „	„ (1)	An ext ∠ of a △ = 2 int opp ∠s	p. 28. „ Cor. 1
13. „	„ 6	Separational Lines are equidistant	p. 32. Th. xxxi. Cor. 2

§ 10. Reynolds.

1. Th   3. Ax

$st_m \mathbb{P} S$ $\overline{X|}$ $|$ $\mathbb{P}$ $\overline{(\ |S|S)'}$

2. Th   4. Th

sS real $S \mathbb{P}. T_m$

6. Th   5. Th

ext $\angle$ of $\triangle =$ $s\ |S|S\ \mathbb{P}\ \overline{S|}\ |$
2 int opp $\llcorner$ s

7. Th   8. Th

$3 \angle$ of $\triangle = 2 \llcorner$ s $X \mathbb{P} T_{d'e''}$

	Tables I–IV		Reynolds
1. Th	I. 5	Different Lines, which make equal $\angle$s with a transversal, are separational	p. 9. Th. VII, and Cors
2. ,,	,, 9	If there be given a Line and a point not on it; a Line can be drawn, through the given point, and separational from the given Line	Not stated: but deducible from above, and assumed as true in Th. x
3. Ax	II. 16(1)	Intersectional Lines cannot both be separational from a third Line	p. 10. Th. VIII. Assumed in course of argument
4. Th	,, 1(2)	Separational Lines make equal $\angle$s with any transversal	p. 10. Th. VIII, and Cors

§ 10. REYNOLDS.

	TABLES I–IV		REYNOLDS
5. Th	II. 15(3)	Different Lines, which are separational from a third Line, are separational from each other	p. 11. Th. IX
6. „	„ 18 (1)	An ext ∠ of a △ = 2 int opp ∠s	„ Th. x
7. „	„ (2)	The 3 ∠s of a △ = 2 ∟s	„ „
8. „	I. 4	An ext ∠ of a △ > either of the 2 int opp ∠s	p. 12. Th. x. Cor. 2

§ 11. WRIGHT.

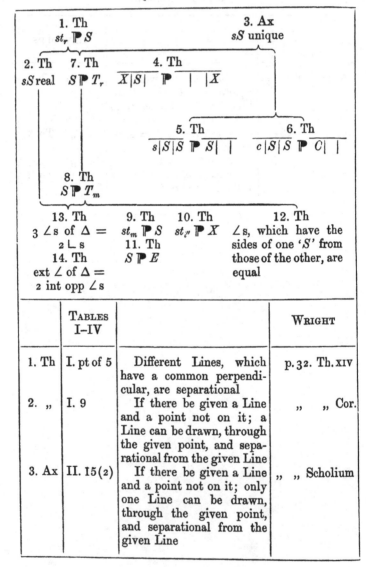

	TABLES I–IV		WRIGHT
1. Th	I. pt of 5	Different Lines, which have a common perpendicular, are separational	p. 32. Th. XIV
2. „	I. 9	If there be given a Line and a point not on it; a Line can be drawn, through the given point, and separational from the given Line	„　„ Cor.
3. Ax	II. 15(2)	If there be given a Line and a point not on it; only one Line can be drawn, through the given point, and separational from the given Line	„　„ Scholium

§ 11. Wright.

	Tables I–IV		Wright
4. Th	II. 16(2)	A Line, which is intersectional with one of two separational Lines, is intersectional with the other	p. 32. § 59
5. „	„ 15(2)	Different Lines, which are separational from a third Line, are separational from each other	„ § 60
6. „	„ „ (3)	Lines, which have a common point and which are separational from a third Line, are coincidental	p. 33. Th. xv. 'This perpendicular being parallel to *AB* can be no other than *CD*.'
7. „	„ 1 pt of (2)	A Line, which is ⊥ one of two separational Lines, is ⊥ the other	p. 33. Th. xv
8. „	„ „ rem of (2)	Separational Lines make equal ∠s with any transversal	„ . Th. xvi, and Cors
9. „	I. 5	Different Lines, which make equal ∠s with a transversal, are separational	p. 34. § 64, 65
10. „	II. 2. pt of (2)	Different Lines, which make with a transversal 2 int ∠s < 2 ∟s, are intersectional	p. 35. § 66. 1°
11. „.	„ 6	Separational Lines are equidistant	p. 36. § 68
12. „	II. 1 (1)	Angles, which have the sides of the one separational from those of the other, each from each, are equal	p. 37. Th. xviii
13. „	„ 18(2)	The 3 ∠s of a Δ = 2 ∟s	p. 41. Th. xix
14. „	„ „ (1)	An ext ∠ of a Δ = the 2 int opp ∠s	„ „ Cor.

§ 12. Wilson's 'Syllabus'-Manual.

1. Th
$X \; \mathbb{P} \; T_{d''e''}$

2. Th   4. Th   3. Ax

$X \; \mathbb{P} \; T_{\iota''}$ $st_a \; \mathbb{P} \; S$ $\overline{X| \;\;|} \; \mathbb{P} \; (\;\; |S|S)'$

5. Th   6. Th   7. Th

sS real $S \; \mathbb{P} \; T_a$ st_a, or st_e, or $st_i \; \mathbb{P} \; t_m$

8. Th   9. Th

$S \; \mathbb{P} \; T_m$ $s|S|S \; \mathbb{P} \; \overline{S| \;\;|}$

10. Th
ext ∠ of Δ = 2 int opp ∠ s

11. Th
3 ∠ s of Δ = 2 ∟ s   12. Th   13. Th
                       sE real $S \; \mathbb{P} \; E$

	Tables I–IV		'Syllabus'-Manual
1. Th	I. pt of 4	An ext ∠ of a Δ > either of the int opp ∠ s	p. 22. Th. 9. Proved as in Euc.
2. „	„ „	Two int ∠ s of a Δ < 2 ∟ s	p. 32. Th. 18. do.
3. Ax	II. 16 (1)	Intersectional Lines cannot both be separational from a third Line	p. 42. Ax. 5. Playfair's Axiom
4. Th	I. 5	Different Lines, which make equal alt ∠ s with a transversal, are separational	p. 43. Th. 21. Proved as in Euc.
5. Th	I. 9	If there be given a Line and a point not on it; a Line can be drawn, through	not stated, but tacitly deduced from Th.

§ 12. Wilson's 'Syllabus'-Manual.

	Tables I–IV		'Syllabus'-Manual
6. Th	II. 1 pt of (2)	the given point, separational from the given Line Separational Lines make equal alt ∠s with any transversal	21 in course of Th. 25 p. 24. Th. 22
7. ,,	not in Tab	Different Lines, which make, with a transversal, equal alt ∠s or equal ext and int opp ∠s or 2 int ∠s = 2 ∟s, make with it equal ∠s	p. 45. Th. 23. Proved by Euc. I. 13, 15
8. ,,	,, 1 (2)	Separational Lines make, with any transversal, equal ext and int opp ∠s, and 2 int ∠s = 2 ∟s	p. 46. Th. 23. Cor
9. ,,	,, 15 (3)	Different Lines, which are separational from a third Line, are separational from each other	p. 46. Th. 24
10. ,,	,, 18 (1)	An ext ∠ of a △ = 2 int opp ∠s	p. 47. Th. 25. Proved as in Euc.
11. ,,	,, (2)	3 ∠s of △ = 2 ∟s	do. do.
12. ,,	,, 8	If there be given a Line and a point not on it; a Line can be drawn, through the given point, equidistant from the given Line	p. 72. ii. Enunciated as a Theorem in words 'Locus of a point, at a given distance &c.' but no proof supplied

§ 12.   Wilson's 'Syllabus'-Manual.

	Tables I–V		'Syllabus'- Manual
13. Th	II. 6	Separational Lines are equidistant	p. 81. Def. 1. Assumed in words '*the* perpendicular distance between the base and the opposite side' but nowhere proved

APPENDIX V.

Proof that, if any one proposition of Table II be granted as an Axiom, the rest can be deduced from it. (See p. 38.)

". . . and so we make it quite a merry-go-rounder." I was obliged to consider a little before I understood what Mr. Peggotty meant by this figure, expressive of a complete circle of intelligence.

It is to be proved that, if any *one* of the propositions of Table II be granted, the rest can be proved.

Euclid I, 1 to 28, is assumed as proved.

It is assumed that, where two propositions are contranominals, so that each can be proved from the other, it is not necessary to include *both* in the series of proofs.

It is assumed as axiomatic that either of two finite magnitudes of the same kind may be multiplied so as to exceed the other.

The following Lemmas are assumed as proved. They may easily be proved by the help of Euc. I. 4, 16, and 26.

LEMMA 1.

If, in two quadrilateral figures, three sides and the included angles of the one be respectively equal to three sides and the included angles of the other, the figures are equal in all respects.

[Analogous to Euc. I. 4.

LEMMA 2.

If, in two quadrilateral figures, three angles and the adjacent sides of the one be respectively equal to three angles and the adjacent sides of the other, the figures are equal in all respects.

[Analogous to Euc. I. 26 (*a*).

T 2

Lemma 3.

It is not possible to draw two perpendiculars to a Line from the same point.

The first cyclical series includes props. 4, 3, 11, 8, 17, 13, 9, 18, each of which is proved by the one next before it, without the help of any other, the last being used to prove the first. Hence, if any one of these eight be granted, the rest can be proved.

(a). II. 4. $\left[st_m \,\rlap{P}\rule[0.7ex]{0.05em}{1.2ex}\, T_m\right]$

Lines, which have a separate point and make equal angles with a certain transversal, do so with all transversals.

Let AB, CD, make equal $\angle$s with EF; and let GH be any other transversal; then shall they make equal $\angle$s with GH.

Join EH.

Then the $\angle$s of the $\triangle\ EFH = 2\ \llcorner$ s; [(θ).

and so also for the $\triangle\ EHG$;

∴ the 4 $\angle$s of $EFHG = 4\ \llcorner$ s;

but the $\angle$s GEF, $EFH = 2\ \llcorner$ s; [*hyp.*

∴ the $\angle$s EGH, $GHF = 2\ \llcorner$ s;

∴ AB, CD, make equal $\angle$s with GH.

Therefore Lines which have &c. Q. E. D.

(β). II. 3. $\left[sT_m \text{ real} \right]$

Through a given point without a given Line a Line can be drawn such that the two Lines make equal angles with all transversals.

For, if through the given point there be drawn a transversal, there can also be drawn through it a Line such that the two Lines make equal ∠s with the transversal;   [Euc. I. 23.

and this Line will be such that the two Lines make equal ∠s with all transversals.   [(α).

Therefore &c.   Q. E. D.

(γ). II. 11. $\left[st_m \mathbb{P} E \right]$

Lines, which make equal angles with a certain transversal are equidistant from each other.

Let *AB*, *CD*, make equal angles with *EF*.

In *CD* take any points *G*, *H*, and draw *GK*, *HL* ⊥ *CD*. Then shall *GK = HL*.

Bisect *GH* at *M*, and draw *MN* ⊥ *CD*.

Now a Line can be drawn through *E* such that it and *CD* make equal ∠s with all transversals;   [(β).

and such a Line would make the ∠ *AEF = EFD* ;

∴ such a Line would coincide with *AB* ;

∴ the ∠ *KNM = NMH =* ∟.

Hence, in the figures *KM*, *LM*, the ∠s at *G*, *M*, *N*, in the one, are respectively equal to the ∠s at *H*, *M*, *N*, in the other, and the sides *GM*, *MN*, in the one, to *HM*, *MN*, in the other;

∴ *GK = HL*.   [Lemma 2.

Therefore Lines &c.   Q. E. D.

(δ). II. 8. [*sE* real]

*Through a given point without a given Line a Line may
be drawn such that the two Lines are equidistant from each other.*

For, if through the given point there be drawn a transversal,
there can also be drawn through it a Line such that the two
Lines make equal ∠s with the transversal;   [Euc. I. 23.
    and this Line will be such that the two Lines are equidistant
from each other.   [(γ).
    Therefore, &c.   Q. E. D.

(ε). II. 17.

*A Line cannot recede from and then approach another; nor
can one approach and then recede from another on the same side
of it.*

If possible, let *ABC* first recede from, and then approach,
DE; that is, let the perpendicular *BG* be > each of the two
perpendiculars *AF*, *CH*.
    From *GB* cut off *GK*=*AF*; and first let *AF*=*CH*.
    Now a Line may be drawn, through *A*, equidistant from
DE;   [(δ).
    and it will pass through *K* and *C*;
    ∴ two Lines, *AKC*, *ABC*, contain a space; which is absurd.
    Secondly, let *AF*≠*CH*; and let *AF* be the greater. Produce
HC to *L*, making *HL*=*AF*.
    Then a Line, drawn through *A*, equidistant from *DE*, will
pass through *K* and *L*, which lie on opposite sides of *BC*;

∴ it will cut BC ;

let it cut it at M ;

∴ two Lines, AKM, ABM, contain a space ; which is absurd.

Similarly it may be proved that ABC cannot first recede from and then approach DE, on the same side of it.

Therefore a Line, &c.   Q. E. D.

(ζ). II. 13. $[se \; \mathbb{P} \; E]$

Lines, of which one contains two points equidistant from the other, are equidistant from each other.

Let AB contain two points E, F, equidistant from CD. From E, F, draw EG, FH, ⊥ CD ; bisect GH in K, and from K draw KL ⊥ CD.

Then $EG=FH$;   [*hyp*.

hence, in the figures EK, FK, the sides EG, GK, KL, in the one, are respectively equal to FH, HK, KL. in the other, and the ∠s at G, K, in the one, to the ∠s at H, K in the other;

∴ the ∠s at L are equal, and are right.   [Lemma 1.

Hence, if the diagram be reversed, KL retaining its position, AB will coincide with BA.

Hence, if there be a point in LA whose distance is < LK, there is another such point in LB, and the Line AB will first recede from and then approach CD ; which is absurd.   [(ϵ).

Similarly if there be one whose distance is > LK.

∴ AB is equidistant from CD.

Similarly, CD is equidistant from AB.

Therefore Lines, &c.   Q. E. D.

(η). II. 9. $\left[se\ \mathbb{P}\ T_m\right]$

Lines, of which one contains two points equidistant from the other, make equal angles with all transversals.

Let AB contain two points equidistant from CD, and let EF be a certain transversal : then shall the $\angle AEF = EFD$.

For AB, CD, are equidistant from each other.   [(ζ).

Bisect EF at G; through G draw $HGK \perp CD$; in CD take two points L, M, equidistant from K; and draw LN, MR, $\perp CD$.

Then $LN = MR$.

Hence, in the figures NK, RK, the sides NL, LK, KH, in the one, are respectively equal to RM, MK, KH, in the other, and the $\angle$s at L, K, in the one, to the $\angle$s at M, K in the other;

∴ the $\angle$s at H are equal, and are right.   [Lemma 1.

Hence, if $\triangle GHE$ be applied to $\triangle GKF$, so that GE coincides with GF, and $\angle EGH$ with the equal $\angle FGK$, EH will fall on FK; otherwise from the same point two perpendiculars would be drawn to GK, which is impossible;   [Lemma 3.

∴ the $\angle GEH = GFK$.

Therefore Lines, &c.   Q. E. D.

(θ). II. 18.

If one side of a triangle be produced, the exterior angle is equal to the two interior and opposite angles; and the three angles of a triangle are together equal to two right angles.

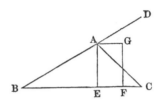

Let *BA*, a side of the triangle *BAC*, be produced to *D*.

From *A* draw *AE* ⊥ *BC*; take *F*, any other point in *BC*, and from *F* draw *FG* ⊥ *BC* and equal to *AE* ; and join *AG*.

Then, ∵ *A*, *G*, are equidistant from *BC*,

∴ *AG*, *BC*, make equal ∠s with all transversals;   [(η).

that is, ∠ *DAG* =*B*, and *GAC*=*C* ;

∴ the whole ∠ *DAC*=*B*+*C*.

To each add ∠ *BAC*;

then the 3 ∠s of the Δ=∠s *DAC*, *BAC*=2 ∟s.

Therefore if one side &c.   Q. E. D.

The second cyclical series includes props. 16 (1), 1 (2), 6.

(κ). II. 16 (1). $[\overline{X|} \quad | \quad \mathbb{P} \; (\quad \overline{|S|S})']$

Intersecting Lines cannot both be separational from the same Line.

If possible, let AB, CD, intersecting at E, be both separational from FG.

In ED take any point K, and draw EH, KLM, $\perp FG$.

Then, $\because AB$ is separational from FG, $EH = LM$;   [(μ).

and $\because CD$ is separational from FG, $EH = KM$;

$\therefore LM = KM$, which is absurd.

Therefore intersecting Lines &c.   Q. E. D.

(λ). II. 1 (2). $[S \; \mathbb{P} \; T_m]$

Separational Lines make equal angles with all transversals.

Let AB, CD, be separational Lines; and let EFG be any transversal.

If $\angle EFB \neq FGD$, let EFB be the greater; and make $\angle EFH = FGD$;

∴ *FH* is separational from *CD* ;   [Euc. I. 26.

∴ the intersectional Lines *AB*, *FH*, are both separational from *CD* ; which is absurd.   [(κ).

∴ ∠ *EFB* is not > *FGD* ;

similarly, it is not less ;

∴ *AB*, *CD*, make equal ∠s with *EFG*.

Therefore separational Lines &c.   Q. E. D.

(μ). II. 6. [*S* ℙ *E*]

Separational Lines are equidistant from each other.

Let *AB*, *CD*, be two separational Lines ; and from any points *E*, *F*, draw *EG*, *FH*, ⊥ to *CD* ; they shall be equal.

Bisect *GH* at *K*, and draw *KL* ⊥ to *CD* ;

∴ the angle *ELK* = *LKH* = a right angle ;   [(λ).

hence, in the figures *EK*, *FK*, the ∠s at *G*, *K*, *L*, in the one, are respectively equal to the ∠s at *H*, *K*, *L*, in the other, and the sides *GK*, *KL*, in the one, to *HK*, *KL*, in the other ;

∴ *EG* = *FH*.   [Lemma 2.

Similarly for any other two points in *AB*.

∴ *AB* is equidistant from *CD*.

Similarly, *CD* is equidistant from *AB*.

Therefore separational Lines &c.   Q. E. D.

It remains to be shewn that, if either cyclical series be proved, a member of the other series can be deduced from it.

This will be done by first deducing prop. 7 (the contra-nominal of prop. 6, a member of the second series) from the first series; and secondly deducing prop. 4 (a member of the first series) from the second series.

(ν). II. 7. [sf 𝕇 X]

Lines, which have a separate point, and one of which has two points unequally distant from the other, are intersectional.

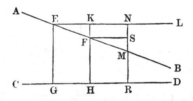

Let AB contain two points E, F, whose distances from CD (EG, FH) are unequal, EG being the greater : they shall be intersectional.

Produce HF to K, making $HK = GE$; join EK, and produce it to L.

Then CD, EL are equidistant from each other,   [(ζ).
and EG is $\perp EL$,   [(η).
and is the common distance between CD and EL.

In FB, produced if necessary, take $FM = EF$; through M draw $NR \perp CD$ and EL; in MN take $MS = FK$, and join FS.

  ∵ KH, NR, are both $\perp CD$,

∴ $\angle EFK = FMS$;   [(α).
also EF, $FK = FM$, MS;

∴ $\angle FSM = EKF$, and is right.

Again, ∵ EL, FS are both ⊥ NS,

∴ they are equidistant;   [(γ).

∴ $NS=KF$;

∴ $NM=$ double KF.

Similarly, by taking in EB, produced if necessary, any multiple of EF, a point will be found in EB whose distance from EL is the same multiple of KF;

but there *is* a multiple of KF which $> EG$;   [Axiom

∴ a point may be found in EB, produced if necessary, whose distance from $EL >$ the common distance between CD and EL;

∴ AB, CD, are intersectional.

Therefore Lines which have &c.   Q. E. D.

(ξ). II. 4. $\left[st_m \ \text{℗} \ T_m \right]$

Lines, which have a separate point and make equal angles with a certain transversal, do so with all transversals.

For they are separational,   [Euc. I. 27, 28.

and therefore make equal $\angle$ s with all transversals.   [(λ).

Therefore Lines &c.   Q. E. D.

APPENDIX VI.

List of propositions of Euc. I, II, with references to their occurrence in the manuals of his Modern Rivals.

§ 1. REFERENCES TO LEGENDRE, CUTHBERTSON, AND 'SYLLABUS'-MANUAL.

EUCLID	LEGENDRE		Page	CUTHBERTSON		Page	'SYLLABUS'-MANUAL		Page
I	**I**		Page	**I**		Page	**I**		Page
Ax 10	Ax	4	6	Ax		2	Def 5		7
11	Th	1	"	Th 8 Cor		17	Th 1		12
12		23	26						
1 Pr				Pr A		11			
2 "				(*p*)		74			
3 "				(*q*)		75			
4 Th	Th	6	10	Th 1		4	Th 5		18
5 "		12	14	2		5	6		20
Cor		Cor	"					Cor	"
6 Th	Th	13	15	4		7	8		22
Cor								Cor	"
7 Th				(*r*)		76			
8 "	Th 11	Sch	14	5		8	15		28
9 Pr	Pr	5	51	Pr B		12	Pr 1		61
10 "		1	49	C		13	4		63
11 "		2	50	7		16	2		62
12 "		3	"	11		20	3		"

§ 1. REFERENCES TO WILSON, PIERCE, AND WILLOCK.

EUCLID	WILSON		PIERCE		WILLOCK	
I	I	Page		Page		Page
Ax 10	Ax 2	3	§16 Th	6		
11	Th	7			Th	10
12						
1 Pr					§ 1	33
2 "					2	34
3 "					3	36
4 Th	Th 16	26	§ 51,52	15	9	46
5 "	8	21	55	16	1	40
Cor	Cor	"	56	17		
6 Th	10	22	58	"	6 (1)	43
Cor	Cor	"	59	"		
7 Th			60	18		
8 "	18	28	61	"	8	45
9 Pr	Pr 1	40	138	39	5	35
10 "	3	42	132	38	6	36
11 "	2	41	134	"	7	"
12 "	4	42	133	"	8	"

Euclid	Legendre	Page	Cuthbertson	Page	'Syllabus'-Manual	Page
13 Th	Th 2	7	Th 9	18	Th 2	13
14 ,,	4	9	10	19	3	14
15 ,,	5	,,	6	14	4	,,
16 (a) ,,			13	22	9	22
(b)			,,	,,	,,	,,
17 ,,			,,	23	18	32
18 ,,	14	15	14	24	10	23
19 ,,	,,	,,	15	25	11	24
20 ,,	8	11	16	26	12	25
21 ,,	9	12	17	27	13	26
	II					
22 Pr			Pr D	30	Pr 5	64
23 ,,	Pr 4	51	E	31	6	65
	I					
24 Th	Th 10	12	Th 18	28	Th 14	27
25 ,,	Sch	13	19	29	16	29
26 (a) ,,	7	11	3	6	7	21
(b) ,,			25 Cor	43	17	30
27 Th	Th 24 Sch	28	Th 20	36	Th 21	43
28 (a) ,,	,,	,,	,,	37		
(b) ,,	22	25	,,	,,	22	44
29 (a) ,,	24 Cor 2	28	21	38	23 Cor	46
(b) ,,	Sch	,,	,,	,,	,,	,,
(c) ,,	24	,,	,,	39	24	,,
30 ,,	25	29	22	40		
	II					
31 Pr	Pr 6	52	Lemma	34	Pr 7	66

EUCLID	WILSON	Page	PIERCE	Page	WILLOCK	Page
13 Th	Th 1	8				
14 ,,	,,	,,	§ 24	8	Th 4	11
15 ,,	,,	9	23	7	2	10
16 (a) ,,	Th 7 Cor4	20				
(b) ,,	,,	,,			12	13
17 ,,					,,	,,
18 ,,	9	21	62	18	6	43
19 ,,	11	22	,,	,,	6 (2)	,,
20 ,,	13	24			7	44
21 ,,	Th	35	40	12		
22 Pr	Pr 5	43	143	40	1	54
23 ,,	5 Cor1	44	136	39	9	37
24 Th	Th 17	27	63	19	12	47
25 ,,	19	29			,,	,,
26 (a) ,,	15	26	54	16	10	46
(b) ,,					11	,,
27 Th						
28 (a) ,,						
(b) ,,						
29 (a) ,,						
(b) ,,						
(c) ,,						
30 ,,						
31 Pr						

EUCLID	LEGENDRE	Page	CUTHBERTSON	Page	'SYLLABUS'-MANUAL	Page
	I					
32 (a) Th	Th 19 Cor 6	23	Th 24	42	Th 25	47
(b) „	19	20	„	„	„	„
Cor 1	20	23	30 Cor	49	26	48
2			30	48	Cor	49
33 Th	30	32	23	41	30	53
34 (a) „	28	30	26	44	27, 28	50
(b) „	„	31	„	„	28	„
					II	
35 Th			Th 31	52	Th 1	82
36 „			Cor	53		
37 „			32 Cor	54		
38 „						
39 „			33 „	55	Th 2 Cor 3	84
40 „			Cor	„	„	„
41 „			32	54		
42 Pr			Pr L	59	Pr 1	99
43 Th			Th 34	56	Th 4	86
44 Pr			Pr M. Cor	60	Pr 2	100
45 „ Cor			N. Cor	61	3	„
46 Pr			Pr K	51		
46 Cor			Th 27	45		
	III		35	57		
47 Th	Th 11	71	36	58	Th 9	91
48 „						

Euclid	Wilson		Pierce		Willock	
		Page		Page		Page
32 (*a*) Th	Th 7	20	§ 71	21	Th 14 Cor 1	15
(*b*) „	„	„	65	20	14	14
Cor 1	6 Cor	16	72	21		
2	6	„				
33 Th						
34 (*a*) „						
(*b*) „						
35 Th						
36 „						
37 „						
38 „						
39 „						
40 „						
41 „						
42 Pr						
43 Th						
44 Pr						
45 „						
Cor						
46 Pr						
46 Cor						
47 Th	29	63	256	74	14	84
48 „					18	88

EUCLID	LEGENDRE		CUTHBERTSON		'SYLLABUS'- MANUAL	
II	**III**	Page	**II**	Page	**II**	Page
1 Th			Th 1	82	Th 5	87
2 ,,			Cor 1	83	Cor 2	88
3 ,,			Cor 2	,,	Cor 1	,,
4 ,,	Th 8	69	Th 2	84	Th 6	,,
Cor						
5 Th					8 Cor	91
Cor	10	70			8	90
6 Th					8 Cor	91
7 ,,	9	,,	2	84	7	89
8 ,,			,,	85		
			I			
9 ,,			Th(*m*)	72	13	96
10 ,,			Cor	,,	,,	,,
			II			
11 Pr			Pr B	89	Pr 6	103
12 Th	Th 13	74	Th 3	86	Th 10	94
13 ,,	12	73	4	87	11	95
14 Pr			Pr A	88	Pr 4	101

Euclid		Wilson		Pierce		Willock	
II			Page		Page		Page
1	Th						
2	,,						
3	,,						
4	,,	Th 25	60			Cor 3	74
	Cor						
5	Th	27 Cor	62			10	82
	Cor	27	61			6	76
6	Th					11	83
7	Th	26	60			5	75
8	,,					9	78
9	,,	28	63			12	83
10	,,					13	84
11	Pr	Ex 3	67			14	104
12	Th	Th 31	66			16	87
13	,,	30	65			17	88
14	Pr	(4)	71	§ 289	84	3	96

§ 2. REFERENCES TO THE OTHER MODERN RIVALS.

EUCLID		CHAUVENET			LOOMIS		
I		**I**		Page	**I**		Page
Ax	10	Ax		12	Ax 11		17
	11	1	Cor	14	1	Cor	18
	12	14	Cor 3	26	23	Cor 3	34
1	Pr						
2	,,						
3	,,						
4	Th	20 Th		30	6 Th		22
5	,,	25 ,,		33	10 ,,		24
	Cor		Cor	34		Cor 2	25
6	Th	27 ,,		,,	11 ,,		,,
	Cor		Cor	,,		Cor	,,
7	Th						
8	,,	22 ,,		31	15 ,,		27
		II			**V**		
9	Pr	28 Pr		78	5 Pr		95
10	,,	25 ,,		76	1 ,,		93
11	,,	26 ,,		77	2 ,,		94
12	,,	27 ,,		,,	3 ,,		,,
		I			**I**		
13	Th	2 Th		15	2 Th		19
14	,,	3 ,,		17	3 ,,		20
15	,,	4 ,,		,,	5 ,,		,,
16(a)	,,						
(b)	,,						
17	,,						
18	,,	26 ,,		34	12 ,,		25
19	,,	28 ,,		35	,, ,,		,,
20	,,	17 ,,		29	8 ,,		23
21(a)	,,				9 ,,		24
(b)	,,	19 ,,		30			

§ **2.** REFERENCES TO THE OTHER MODERN RIVALS.

EUCLID	MORELL	Page	REYNOLDS	Page	WRIGHT	Page
I	**I**		**I**		**I**	
Ax 10	Ax	1			Ax	1
11	Th 1 Cor	5	Ax	2	Th	6
12	„ 25 Cor	26			„ 16 Sch	35
1 Pr						
2 „						
3 „						
4 Th	Th 5	9	Th 14	15	Th 8 (2)	16
5 „	, 10	13	„ 11	12	„ 9 (3)	19
Cor	Cor 1	,,	Cor 2	,,	Cor	20
6 Th	„ 11	14	„ 12 Cor	13	„ 9 (1)	18
Cor	Cor	15			Cor	20
7 Th						
8 „	„ 9	12	„ 13	14	„ 8 (3)	16
					II	
9 Pr	Pr 9	66	Pr 2	30	Pr 9	87
10 „			„ 1	29	„ 8	86
11 „			„ 3	31	„ 11	88
12 „	„ 7	64	„ „	,,	„ „	,,
13 Th	Th 2	6	Th 1	3	Th 2	8
14 „	„ 3	7	„ „	,,	„ 3	,,
15 „	„ 4	8	„ 2	4	„ 5	11
16 (a) „			„ 10 Cor 2	12		
(b) „			„	,,		
17 „						
18 „	„ 12	15	„ 12	13	„ 9 (4)	19
19 „	„ „	,,	„ „	,,	„ 9 (2)	18
20 „			„ 3	5	„	13
21 (a) „	„ 8	12	„ 4	,,	„ 6	14
(b) „						

EUCLID	CHAUVENET	Page	LOOMIS	Page
	II		**V**	
22 Pr	34 Pr	81		
23 ,,	29 ,,	79	4 Pr	95
	I		**I**	
24 Th	24 Th	33	13 Th	26
25 ,,	Cor	,,	14 ,,	27
26 (a) ,,	21 ,,	31	7 ,,	23
(b) ,,				
27 ,,	14 ,,	25	22 ,,	33
28 (a) ,,	Cor 1	,,	,, ,,	,,
(b) ,,	,, 2	26	21 ,,	32
29 (a) ,,	13 ,,	24	23 ,,	33
(b) ,,	Cor 2	25		
(c) ,,	,, 3	,,		
30 ,,	12 ,,	24	24 ,,	34
	II		**V**	
31 Pr	30 Pr	79	6 Pr	96
	I		**I**	
32 (a) Th	18 Cor 1	29	27 Th	36
(b) ,,	,, Th	,,	,, ,,	,,
Cor 1	29 ,,	37	28 ,,	37
,, 2	Cor 2	38	29 ,,	,,
33 Th	32 ,,	40	32 ,,	,,
34 ,,	30 ,,	39	30 ,,	38
35 ,,				
36 ,,				
37 ,,				
38 ,,				
39 ,,				
40 ,,				
41 ,,				

EUCLID		MORELL		REYNOLDS		WRIGHT	
			Page		Page	**II**	Page
22	Pr	Pr 6	64	Pr 7	32	Pr 6	83
23	,,	,, 1	61	,, 4	32	,, 1	79
24	Th	Th 7	10			Th 7	15
25	,,						
26 (a)	,,	,, 6	10	Th 15	16	,, 8 (1)	16
(b)	,,						
27	,,	,, 25	25	,, 7 Cor 2	9	,, 16	33
28 (a)	,,	,, ,,	,,	,, 7	,,	,, ,,	,,
(b)	,,	,, ,,	,,	Cor 3	,,	,, ,,	,,
29 (a)	,,	,, 24	24	,, 8 Cor 2	10	,, ,,	,,
(b)	,,	,, 23	22	,, 8	,,	,, ,,	,,
(c)	,,	- ,, 24	24	Cor 3	,,	,, ,,	,,
30	,,	,, 22	22	,, 9	11	,, 14 Sch	60
						II	
31	Pr	Pr 8	65			Pr 7	85
32 (a)	Th	Th 28 Cor 1	28	Th 10	11	Th 19 Cor	41
(b)	,,	,, 28	27	,, ,,	,,	,, 19	,,
Cor 1		,, 29	29	,, 18	20	,, 20	42
2		,, 30	30	,, ,,	,,	Cor	43
33	Th	,, 33	33	,, 17	20	,, 22	48
34	,,	,, 31	31	,, 16	19	,, 21	47
				and III 1	60		
				III		**IV**	
35	,,			Th 2	61	Th 1	185
36	,,			Cor	,,		
37	,,			,, 4	62	Cor	186
38	,,			Cor 1	,,		
39	,,			Cor 2	63	Sch	187
40	,,			,,	,,		
41	,,			,, 3	62		

Euclid	Chauvenet		Page	Loomis		Page
42 Pr						
43 Th						
44 Pr						
45 ,,						
Cor						
46 Pr						
	IV			**IV**		
46 Cor						
47 Th	10	Th	133	11	Th	73
48 ,,						
II						
1 Th						
2 ,,						
3 ,,						
4 ,,				8	Th	71
Cor						
5 Th						
Cor				10	,,	73
6 Th						
7 ,,				9	,,	72
8 ,,						
9 ,,						
10 ,,						
11 Pr						
12 Th	16	Th	113	13	Th	76
13 ,,	15	,,	112	12	,,	75
14 Pr	12	Pr	136	22	Pr	103

EUCLID	MORELL	Page	REYNOLDS	Page	WRIGHT	Page
42 Pr			Pr 1	76	Pr 2	198
43 Th			Th 1 Cor	60		
44 Pr						
45 ,,						
Cor						
46 ,,						
	V				**IV**	
46 Cor						
47 Th	Th 8	149	Th 12	69	Th 2	187
48 ,,						
II					**IV**	
1 Th						
2 ,,						
3 ,,						
4 ,,	Th 9	151	Th 9	67	Th 4	190
Cor						
5 Th					5	191
Cor	,, 11	152	,, 11	68	6 Cor	193
6 Th					,, 6	192
7 ,,	,, 10	152	,, 10	67	4 Sch	191
8 ,,						
9 ,,						
10 ,,						
11 Pr					Pr 3	198
	III		**IV**			
12 Th	Th 22	101	Th 17	97	Th 7	193
13 ,,	,, 21	100	,, 16	,,	8	194
14 Pr					Pr 2 Sch	198

Printed in the United States
By Bookmasters